DEATH OF AN OCEAN

DEATH OF AN OCEAN
A Geological Borders Ballad

Euan Clarkson

and

Brian Upton

DUNEDIN

Published by
Dunedin Academic Press Ltd
Hudson House
8 Albany Street
Edinburgh EH1 3QB
Scotland

ISBN 9781906716028

British Library Cataloguing in Publication Data
A catalogue record for this book is available from the British Library

Prepress design and production
by Makar Publishing Production, Edinburgh
Diagrams by Jacque Barber – www.palaeo-art.com
Printed and bound in Poland, produced by Hussarbooks

Contents

List of Abbreviations and Illustrations

Preface

Our earlier book, *Edinburgh Rock*, was published by Dunedin Press in 2006, following on from Brian Upton's *Volcanoes and the making of Scotland* of 2004. Whereas some aspects of the geology of the Scottish Borders have been discussed in these works, it seemed to us that there was much more that we could have written on this beautiful and geologically fascinating corner of Scotland. Accordingly, we began work on the present volume, visiting many new locations, as well as those to which we had taken student parties over the past forty years.

The area which we cover in this book is approximately coincident with, though somewhat more extensive than, the politically defined 'Borders' Region and generally lies east of the River Clyde valley and the M74 motorway. Our title *Death of an Ocean* refers to the ancient ocean, known as Iapetus, within which much of the Borders rocks originated. This ocean once lay between what is now the Scottish Highlands, itself once part of the Laurentian Continent (mainly North America and Greenland), and the smaller southern continent of Avalonia (now distributed between eastern Newfoundland, southern Ireland, Wales and England).

It is, however, not with the ocean's growth but its demise, as recorded in the Southern Uplands of the Borders, that this book is principally concerned. Its closure was brought about by the oblique collision of the Laurentian and Avalonian continents, in the early Palaeozoic, more than 400 million years ago. Sediments deposited on the ocean floor and converted over time into sandstones and shales, were buckled and broken as they became trapped between the impacting continents. The closure process involved the formation not only of profoundly deep ocean trenches but the growth of accompanying chains of volcanoes. Our account includes descriptions of the ecosystems of the ocean. In particular, it deals with the oceanic plankton, mainly in the form of fossil graptolites, and shows how the Palaeozoic plankton differed from that of modern oceans.

The book explains how, in terms of modern plate tectonic theory, oceans can be created and destroyed, how the idea of Iapetus was first conceived and how geological research has convincingly demonstrated its former reality. Whereas the rocks representing the Iapetus Ocean floor are best exposed along the Berwickshire coast, they can also be studied in innumerable river sections, road-cuttings and quarries.

We would remain wholly ignorant of the astonishing and dramatic secrets of this ancient Borders saga were it not for the diligent contributions of a host of men and women over the past two hundred years. Through their intellectual curiosity, backed by countless hours of field-work (much of it under harsh conditions), as well as laboratory studies, the story has gradually emerged. Accordingly, we pay due respect to some of the indefatigable individuals who have devoted large parts of their working lives to unravelling the saga encrypted in the rocks, fossils and landforms of the Borders. It is, however, to Stuart McKerrow, who was both our friend and teacher, that the book is dedicated. His unquenchable enthusiasm stimulated much of the intensive investigations over the past forty years. Our book is neither a comprehensive field-guide nor a treatise for the professional geologist but an attempt to tell the story of Iapetus, as well as its later Palaeozoic aftermath, for the non-specialist reader.

It was following the ocean closure and the extreme deformation and mountain-uplift that went with it that the future 'Borders' began to emerge. In its early stages it presented a volcanic, desert landscape that gradually saw a 'greening' as plants began to gain a tentative foot-hold on dry-land. Subsequently, as rainfall and temperatures rose during what is called the Carboniferous Period, the putative 'Borders' became increasingly lushly vegetated. Several dozen volcanoes came into existence and it is the deeply eroded, sawn-down stumps left by these that form many of the most scenic features of the Borders. Although in early Carboniferous times (roughly 340 million years ago) the land, rivers and lakes would have been teeming with invertebrate animals as well as fish, amphibians and primitive reptiles, we are still at a stage in Earth history that was roughly a hundred million years before the first dinosaurs arrived upon the scene. The fact is that, for better or worse, the geological record in the Borders virtually came to a close and there is extremely little evidence for any significant events younger than the early Carboniferous until the onset of the Pleistocene ice-ages in the recent geological past. It is, with the modifications of the landscape bequeathed by these Pleistocene glaciations that we conclude the book.

We gratefully acknowledge the help and enthusiasm of our colleagues who have shared their geological insights with us and contributed greatly to our own understanding. Special thanks go to Gordon Craig and Ken Walton, our mentors in the early 1960s, who introduced us to the Borders geology. We wish to thank Cecilia Taylor, David Harper, Alan Owen, Colin Scrutton, Howard Armstrong, Neil Clark, John Cater, Lyall Anderson, Sarah Stewart, Yves Candela, Adrian Rushton, Philip Stone, John Dewey, Roger Hipkin, Jim Floyd, Brian Bluck, Ray Macdonald, Alastair

Robertson, John Craven, David Stevenson, Andrew Macmillan and many others. Our thanks too to Bill Crighton for his excellent photographs of graptolites, displayed in Chapter 9, and David Loydell for ensuring that the graptolite names used herein are up to date, and we have enjoyed stimulating discussions with Eva Panagiotokalu. The maps and field sketches were redrawn from our originals or from other sources by Jacque Barber. We are much indebted to Yvonne Cooper and to Bill Gilmour for their unstinting and enthusiastic help with respect to obtaining the photographic illustrations. Finally, we express our gratitude to David McLeod and Anne Morton for their support and encouragement during the editing and production of this book.

CHAPTER 1

Introduction

The ocean in question

Whilst much has been written concerning the history of the Borders from Bronze Age times to the present, it is our intention to write a story about its more ancient prehistory – the story as recorded in its rocks and landscape.

So what is this ocean that has died? In these days of global warming and environmental pollution, it might well be thought that this book concerns an ocean so horrendously polluted that all life in it has been destroyed. This however, is not what we write about. Rather this book concerns the evidence pointing to the former existence of a great ocean named after the Greek god, Iapetus. Iapetus was a Titan, born to Uranus and Gaia. He fathered Atlas, Prometheus, Epimetheus and Menoetius by an Oceanid called Clymene and, through the Atlas line, he was considered to be an ancestor of the human race.

Early suspicions that such an ocean had actually existed arose primarily from geological studies in the north of England and southern Scotland. We, however, are going to focus attention specifically on the geology of south-eastern Scotland, specifically the Borders area, lying between the Scottish Midland Valley to the north and England to the south. Our story is mostly devoted to events that took place in the Palaeozoic Era, the stretch of geological time currently defined as between 542 and 299 million years ago (hereafter, Ma). Palaeozoic means 'old life' and is so named because the fossils preserved in rocks of this age are very significantly different from those of the two succeeding eras, namely the Mesozoic ('middle life') and the Cenozoic ('recent life'). Some of the fossils, especially from the older (or Lower) Palaeozoic rocks, bear no clear relationship to any living plants or animals. The majority, however, clearly indicate some sort of kinship with modern forms. Thus the trilobites had much in common with present-day 'jointed-legs' invertebrates like crustaceans, insects and spiders. The Palaeozoic rocks contain fossils that are recognisable versions of starfish, shell-fish, corals and plants. Towards the close of the era, fish, amphibians and reptiles had acquired forms that can readily be accepted as ancestral to those with which we are familiar.

The Iapetus Ocean is believed to have been created some six hundred million years ago when a great continent broke up in response to stresses exerted by the motion of the hot mantle rocks beneath it. It is this persistent convective turnover in the mantle that pulls and pushes the tectonic plates that constitute the outer part of the solid Earth and, as a consequence of further plate movements, the ocean closed again as the continents bounding it converged and collided. Figuratively, we can say that the Iapetus Ocean was 'born' at about 616 Ma and finally 'died' some two hundred million years later at approximately 420 Ma. If 616 Ma seems a long time ago, it is worth reflecting that, since the Earth is approximately 4600 m.y. old, some seven-eighths of Earth-time had already elapsed before the break-up.

The oldest rocks in the Borders are those that compose the Southern Uplands: these rocks were derived from sands, muds (including 'oozes') and some volcanic lavas that accumulated on the deep ocean floor. As plate tectonics drew the neighbouring continents ever closer, the soft sedimentary materials were converted into hard rocks (sandstones, shales and cherts). These, together with the (very subsidiary) lavas, became folded, broken and uplifted in a long episode of mountain building (an orogeny) attending the continental collision.

Understanding how a continent can break into pieces and how oceans start and end, as well as how fold-mountain belts are generated, only became possible after the formulation of the theory of plate tectonics, less than fifty years ago. In this book we try to explain, with the least recourse to technical terminology, how the existence of this old Iapetus Ocean was first suspected and later proved beyond all reasonable doubt. The rocks of the Borders played a key role in providing the evidence. Just as it took generations of scholars to decipher the ancient Egyptian hieroglyphs of the Rosetta Stone, so extracting the secrets of Iapetus from the rocks of the Southern Uplands has taken the efforts of many men and women over the past two centuries.

As explained above, Iapetus had a 'lifespan' of nearly two hundred million years and, with its closure, a new large continent, called Laurussia, came into being. The mountains constructed out of the highly deformed oceanic sediments started to undergo erosion from the very beginning of their emergence above sea-level. The erosive products were carried away and deposited and a new cycle of sediment deposition began. The younger sedimentary rocks that formed from this deposition, together with some volcanic products, largely covered the eroded remnants of Iapetus. These Upper Palaeozoic rocks, dating from around 416 to 350 Ma, overlie and cover up much of the older Palaeozoic rocks from which the history of Iapetus has been gleaned. It is the sedimentary rocks of the upper Palaeozoic that underlie the fertile lands of the Borders. And it is the associated igneous rocks allied to various phases of volcanism that are responsible for most of the scenic features of the Borders such as the Eildon Hills above Melrose and the desolate moors of the Cheviot.

Let us return to the subject of Iapetus and its status as an ocean, and consider how an ocean differs from a sea. While it is always difficult to make absolute definitions, the most obvious point is that oceans tend to be larger and deeper. Less obviously, they are underlain by crust which has a composition and geological structure significantly different from that beneath seas. Although oceans are usually bigger, we shall show that oceans grow from small ('embryonic') beginnings and terminate by shrinkage and ultimate closure. Seas are rarely over 1 km deep and are often much shallower. Oceans, on the other hand, not only have deep floors ('abyssal plains') lying 3 to 4 km below the surface, but can contain profound trenches that descend to depths of over 10 km.

Whereas the crust beneath the deep oceans has a composition and structure quite distinct from those of continents and seas, their marginal parts – the continental shelves – generally have a make-up similar to that of the continents themselves. So oceans tend to be big and deep and are underlain by rocks materially different from those below mere seas. Seas, being shallower, can be affected by changes of sea-level to the extent that they can dry up and disappear entirely during times of exceptionally low global sea-level. Thus only a few thousand years ago it was possible to walk across what is now the North Sea to mainland Europe. The Mediterranean has had a complex history of emptying and refilling and the Black Sea may only exist as a result of dramatic infilling consequent upon an exceptional sea-level 'high' in the Mediterranean. But whilst the rise and fall in sea-levels can obviously change the geography around the edges of an ocean, the fundamental nature of the ocean floor geology remains constant. It is primarily the restless motions of the tectonic plates that dictate where an ocean will develop, how large it will grow and how long it will exist.

Admittedly it takes a great leap of imagination when looking at the rounded hills of the Southern Uplands to think that these are the mute memorials to a once great ocean. The sands, silts and muds that went into the construction of the Southern Uplands were deposited, not on the floors of any shallow and relatively insignificant seas but in the black depths of an ocean which, in its heyday, would have rivalled any of the great oceans of the present day.

Our book gives an account of this ocean, not in its glory days but in the later times that led to its erasure. We try to explain the nature of the sediments and the manner in which they were transported from landscapes now long gone and deposited in the

1.1 The Moorfoot Hills, view west from the B709, south of Middleton.

ocean deeps. And of how they were then transmuted into rocks which, in turn, were to suffer extreme compression and deformation. The account tells of the myriad small organisms that lived in the surface waters of the ocean and which, on dying, added their remains to the oozes on the ocean floor. It discusses also the evidence for the volcanic archipelagos that were intimately associated with the ocean closure.

Although the ocean grew and shrank in the Lower Palaeozoic, this was, as noted, only a short time ago relative to the overall history of our planet. Nonetheless, they were times long pre-dating not merely the dinosaurs but all vertebrate life on land. Fishes swam in the Iapetus Ocean and it is sobering to realise that in those Palaeozoic days our own ancestors were to be found among them.

Hills and dales

The Southern Uplands comprise gently rounded hills which lack the dramatic topography of the Scottish Highlands and just fail to reach a height of 850 m. To the east, the Border country is bounded by the widening estuary of the river Forth and the North Sea. Its northern marches are defined by the Lammermuir and Moorfoot Hills (Fig. 1.1), the escarpments of which form the southern horizon as viewed from the vicinity of Edinburgh. Away to the south-east lies the great hulk of the Cheviot, rising

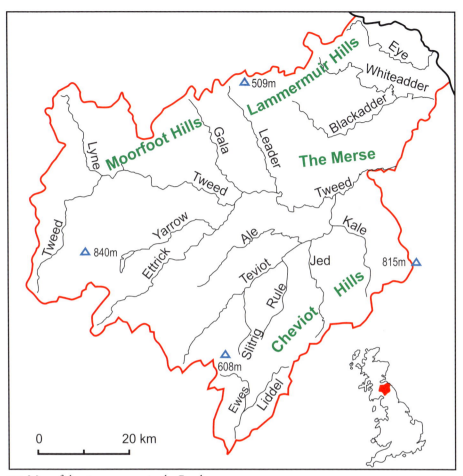

1.2 Map of the river systems in the Borders.

to 815 m. The Cheviot straddles the Scottish–English boundary and only its smaller portion lies in Scotland. We shall take as our western boundary the old north–south route along Clydesdale and Annandale utilised by the 'West Coast' railway and the M74 motorway linking Glasgow and Carlisle.

The river Tweed runs through the heart of the Borders (Fig. 1.2) flowing past Peebles, Galashiels, Melrose, Kelso and Coldstream, to reach the sea at Berwick-upon-Tweed. The broad lower part of the Tweed valley is known as the Merse. The principal tributaries of the Tweed are the Leader Water, Gala Water, Yarrow, Ettrick and Teviot. Further to the north-east, the Blackadder Water combines with the Whiteadder Water and these, like the Eye Water, flow into the North Sea. At the extreme south end of the Borders and on the other side of the watershed, the Ewes Water joins the Liddel Water to flow southwards past Longtown, to the Solway Firth (Fig. 1.2).

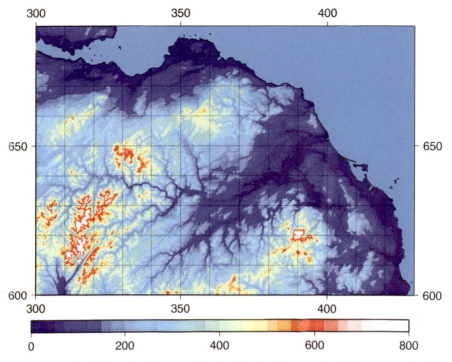

1.3 Topographic (false-colour) image of SE Scotland and NE England drawn using data from the Shuttle Radar Topography Mission. The image covers the Borders and shows the coastlines from near Grangemouth in the NW to near Alnwick in the SE. Highground (yellow/orange) centre left is the Southern Uplands and Cheviot appears lower right. The dendritic pattern of the Tweed and its tributaries is well displayed.

The Scottish Borders have inspired notable writers such as James Hogg ('the Ettrick Shepherd') and Sir Walter Scott. Charles Lapworth was an Oxfordshire teacher who was to play a critical part in unravelling the geological history of this land. He first came to live and work in the Borders because he had been inspired by Scott's novels. So, although he never knew it, Walter Scott indirectly had an influence on the development of geology.

CHAPTER 2

The Time Lords

An essential precursor to our understanding of the nature of our planet was the appreciation that the Earth orbits the sun rather than vice versa. The revolutionary ideas of Copernicus were subsequently verified by Galileo and Kepler in the late sixteenth to early seventeenth centuries, following the invention of the telescope. Another major step in the appreciation of our status in the universe came late in the eighteenth century when the immensity of geological time was first demonstrated. The name most closely associated with recognising the huge passages of time involved in Earth history is that of James Hutton (1726–97) a man, incidentally, whose life was very much bound up in the Scottish Borders.

Archbishop Ussher arrived at a very precise age for the Earth in the seventeenth century. Computing from biblical records, he concluded that the Earth began on 23 October 4004 BC. Although widely accepted, it was not undisputed: Isaac Newton had concluded that if the Earth had commenced as a red-hot sphere, it would have taken some 50,000 years to cool down.

In some other cultures that had not been straitjacketed by the time limitations imposed by literal acceptance of the Bible, it had long been accepted that the Earth was of great antiquity. Indeed, Indian cosmologists were the first to estimate the age of the Earth as being over two billion years and perhaps more than four billion years. An ancient Hindu legend that beautifully illustrates the concept of the enormity of geological time tells of an immortal eagle that flies over the Himalayas once every thousand years. The eagle carries a feather in its beak and, every time it passes across, it gently brushes the summits of the great mountains with its feather. The length of time that it takes the eagle to completely erode the mighty Himalayas was said to be the age of the universe.

Nonetheless, in the West it took the deductive reasoning of James Hutton to bring geological timescales into focus and to allow the great intellectual leap essential to

2.1 Hutton's unconformity at Siccar Point, where gently dipping Upper Old Red Sandstone overlies vertical, eroded Silurian strata.

geological interpretation. In 1750 Hutton inherited his father's farm in Berwickshire. As a farmer he realised that soils, as well as the sands and gravels of the rivers and beaches, all come from the breakdown of solid rock, through the agency of water, wind or ice. And yet, since not all the Earth's surface had so been degraded, he argued that there must be some complementary processes for its renewal. He travelled widely

throughout Britain and, in so doing, he developed a passionate interest in rocks. Near Jedburgh, he noted that red sandstone strata lay approximately horizontally abruptly above dark grey sandstones that stood nearly vertical. The name 'greywacke' is now universally employed to describe such dark grey sandstones (of which more later!) but the name Hutton used for them was 'schistus'. Hutton considered the implications of the contrasting attitudes of the two suites of rock. He and his friends subsequently described another place where the same phenomenon, subsequently to become known as an 'angular unconformity', could be observed. This now widely celebrated locality is Siccar Point on the East Lothian coast (Fig. 2.1). Hutton's meticulous logic took him to the conclusion that the steeply-dipping greywacke ('schistus') strata must themselves have been laid down as essentially horizontal layers of sand that were subsequently rotated nearly ninety degrees by some great forces of the Earth. Furthermore, he postulated that the 'schistus' had then been truncated by erosion to produce a ragged sub-horizontal surface. Lying at low angles upon this are red-brown sandstone strata, comparable to the situation at Jedburgh.

Siccar Point is now revered as 'Hutton's unconformity' and has become a focus for geological pilgrimage. Hutton reasoned that the time taken for accumulation of the greywacke sediment at the bottom of some long-forgotten sea must have been measurable in many thousands of years. Further eons would firstly have been required for their hardening into rock, secondly for operation of the forces that rotated them and thirdly, for their uplift and exposure to the weathering processes that created the ragged surface. And after all that, further immense periods of time would have had to elapse to allow for the deposition of the overlying red sandstones, which themselves were then tilted, uplifted and partially eroded. Such thoughts led Hutton to the realisation that, rather than the few thousand years prescribed by religious orthodoxy, uncounted millions of years must have been required to account for the rock sequence at Siccar Point. By these arguments the case was incontrovertibly established that geological processes proceed over hitherto unimaginably long spans of time. Hutton's fame rests on his discovery of what is now commonly called 'deep time' and his recognition that the Earth behaves as a nearly eternal machine. While mountains are constantly being eroded and the products transported by rivers for deposition in the seas, other processes are simultaneously uplifting new mountain ranges. His ideas were enthusiastically adopted by Charles Lyell (1797–1875) who encapsulated them in the dictum that '*the present is the key to the past*'.

Although there were no means of quantifying geological time in the eighteenth and nineteenth centuries, the discovery of radioactivity towards the end of the nineteenth century led in the twentieth and twenty-first centuries to the development of highly sophisticated and precise means of dating rocks. A name of the greatest importance in the early development of radiometric dating was that of Arthur Holmes, a genius who successively held the position of Professor of Geology at Durham and Edinburgh Universities. James Hutton and Arthur Holmes have to be acknowledged as the two great Time Lords (with all due respect to Doctor Who!). Radiometric dating is based on the fact that, during the formation of rocks (by compaction and cementation of the grains in the case of sediments, or the crystallisation of molten magma in the case of igneous rocks), various radioactive elements ('unstable isotopes') become incorporated.

Among the elements most useful for radiometric dating are uranium, rubidium and potassium. The natural decay of the various unstable isotopes results in generation of new or 'daughter' isotopes. By determining the rate at which the different radioactive isotopes break down to create daughter products, and by measuring the amount of daughter isotopes so formed, the time of rock formation can be ascertained.

The concept of radiometric dating, which would surely have fascinated Hutton and baffled Archbishop Ussher, has become a very refined science. It has permitted its practitioners to fix the time that the Earth was created at approximately 4.6 thousand million years. Furthermore, we now know that most of the rocks composing the Borders were formed over approximately one hundred million years within four of the geological periods that compose the Palaeozoic Era. These are, from oldest to youngest, the Ordovician (488–444 Ma), Silurian (444–416 Ma), Devonian (416–359 Ma) and Carboniferous (359–299 Ma) Periods. Rocks dating from the earliest Period of the Palaeozoic, the Cambrian (542–488 Ma), are unrepresented in the Borders, the oldest rocks exposed being Ordovician, no older than 470 Ma. The latest Period of the Palaeozoic Period was the Permian (299–251 Ma). Rocks of this age occur in Annandale, just west of the region considered in this book.

CHAPTER 3

Geological overview of the Borders

As explained, the bedrocks of the Borders embrace four of the geological periods that compose the Palaeozoic Era, namely the Ordovician, Silurian, Devonian and Carboniferous Periods. The Ordovician and Silurian Periods were both defined by geologists working in Wales in the nineteenth century, the names deriving from two of the old Celtic tribes of that country, the Ordovices and the Silures. The Devonian and Carboniferous Periods also gained their names in the nineteenth century; the one from researches in south-west England and the other because the principal economic coals were laid down in that time.

The Lower Palaeozoic

Ordovician and Silurian rocks crop out in a broad SW–NE swathe across southern Scotland to form the Southern Uplands (Fig. 3.1). The outcrop area is broadest in the south-west (some 65 km), narrowing north-east to around 20 km across in the Borders. Silurian rocks dominate, with only a subordinate strip of Ordovician cropping out along the northern margin. The Ordovician rocks underlie the Moorfoot and Lammermuir Hills with their faulted northern escarpment dropping down to the softer Upper Palaeozoic strata of the East Lothian lowlands.

It has to be admitted that, at first acquaintance, the bulk of the rocks forming the Southern Uplands do not look exciting. Along the M74 the traveller sees nothing but a series of monotonous sandstones in the many road cuttings. These are marine sandstones of the 'greywacke variety' which figure large in our account of the closure of the Iapetus Ocean. So, at first sight a more boring and uninspiring set of rocks could hardly be imagined but, as we shall see, they encapsulate an enormous quantity of historical and environmental data (Fig. 3.2).

3.1 View east to the Southern Upland hills (Fingland Rig and Pykestone Hill near Drumelzier).

The Caledonian orogeny arose in a series of uplifts towards the end of the Silurian Period to form a mountain chain thousands of kilometres long. At its maximum development this would have rivalled the Alps and possibly the Himalayas, in length, height and grandeur. The creation of this great Caledonian mountain range was the culminating event that marked the final closure of the Iapetus Ocean. It was the gravestone above the defunct ocean.

The Upper Palaeozoic

As the mountains were worn away, the rivers draining them evolved from fierce mountain torrents to tamer, sluggish streams. In consequence, the early erosive stages following orogenesis typically yielded coarser sediments, whereas in the later, more mature stages, the sediments become finer and finer. Consequently, conglomerates and coarse-grained sandstones were characteristic of the early post-orogenic stages and compose what is known as the 'Lower Old Red Sandstone' succession. This name was proposed in 1822 by Conybeare and Phillips to describe the mainly river and lake deposits laid down in the aftermath of the Caledonian Orogeny. Such deposits are referred to as 'terrestrial' in contrast to marine sediments laid down off-shore. The reason for calling it a 'sandstone formation' is obvious enough, while the 'red' derives from the almost ubiquitous presence of the red iron oxide, haematite, precipitated by percolating waters in the oxidising desert conditions. Comparable red sandstones were typical products of the much younger desert environments in the Permian and Triassic Periods, and the terms 'Old' and 'New' were used to distinguish these two predominantly red sandstone formations.

See page 129

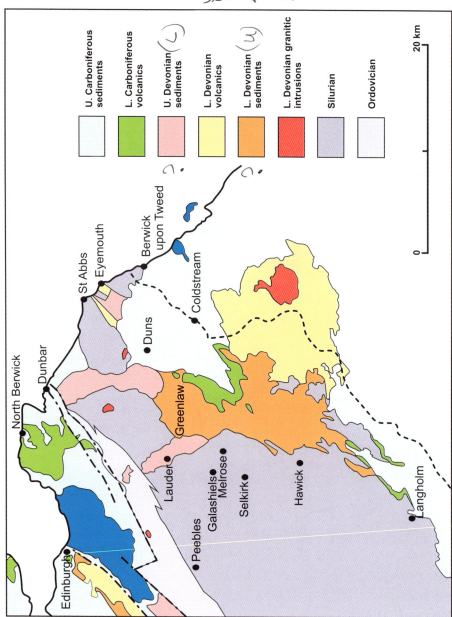

3.2 Geological map of the Borders.

The Scottish Old Red Sandstone rocks were later subdivided by Sir Roderick Murchison into three, namely Lower, Middle and Upper sequences, approximately corresponding to products of Lower, Middle and Upper Devonian times. In the Borders region, however, whereas the Lower and Upper sequences are represented, Middle Old Red Sandstone rocks are absent. Although Lower Old Red Sandstone times saw an abundance of volcanoes away to the north, in what were to become the Midland Valley and Grampian Terranes, it was only rather later that eruptions took place in the Borders.

Some time after 390 Ma tectonic disturbances resumed: these were pale echoes of distant continental plate collisions far to the south, causing much gentler folding than that which preceded the Lower Old Red Sandstone deposition. A prolonged period of erosion during mid-Devonian times removed much of the Lower Old Red Sandstone products, accounting for their scattered outcrops. This Middle Devonian erosion was followed by further deposition of sandstones (principally by rivers), to produce the Upper Old Red Sandstone succession. Outcrops of these occupy an irregular (but crudely north–south trending) stretch of the Borders from near Polwarth and Westruther south towards Hawick and Bonchester. They can be seen near Earlston and Duns and upon the Cheviot lavas in parts of Roxburghshire.

Whereas in some places the Upper Old Red Sandstones can be seen lying with marked discontinuity above the Lower Old Red deposits, elsewhere (e.g. between Earlston and Hawick) the latter were completely removed by erosion in the Middle Devonian, and the Upper Old Red Sandstones rest directly upon the steeply folded early Silurian strata.

Some hundred million years passed after the climax of the Caledonian Orogeny before the Carboniferous Period commenced. By this stage the mountains had been extensively reduced and the sediments coming from them were typically finer-grained than they had been in the preceding Devonian Period. Much of the Borders region had an elevation not much above sea-level. The early Carboniferous sediments were largely deposited by rivers or lakes, but with increases in global sea-levels, the lowlands were periodically inundated by shallow seas.

Plate movements, although concentrated several hundred kilometres to the south, still had repercussions in northern England and southern Scotland. These involved some stretching (extension) of the crustal rocks with the generation of faults. Stretching led to reduction of pressure on the underlying hot mantle, triggering magma generation and a resumption of volcanism from the northern shores of the Solway Firth to the Borders. The lavas from these volcanoes, together with the worn-down stumps of the sub-volcanic intrusive rocks, give rise to many of the most scenic features in the Borders. Following this and two more brief volcanic episodes, parts of the Borders continued to subside and receive river-borne sediments from the now much reduced mountains to the north-west. Thus much of the land beneath, for instance, Coldstream, Duns and Berwick-upon-Tweed became overlain by fluviatile sandstones deposited in the early phases of the Carboniferous Period.

Post-Palaeozoic geology of the Borders

From about 340 Ma when the sedimentary and volcanic rocks of the Lower Carboniferous were formed to the present day, there was remarkably little geological change in the Borders. However, in comparatively recent times, temperatures fell dramatically in the Pleistocene Period at 1 to 2 Ma and massive ice-sheets developed across Scotland. The movement of these glacial sheets scoured and carved the landscape of the Borders essentially into its present morphology whilst their melting left a blanket of morainal material across the area. Torrential meltwaters from the decaying ice-sheets cut deep into existing valleys. These Pleistocene modifications are discussed in the final chapter.

Borders geology in relation to the rest of Scotland

Before focusing on the Borders we need briefly to consider the geology of Scotland as a whole. Among the most striking features of Scottish geology are the great faults that define the boundaries of the Midland Valley. These are the Highland Boundary Fault and the Southern Upland Fault, running ENE–WSW and almost parallel. A fault is a plane (becoming a line on a map!) where the rocks on one side have been moved against those of the other: the displacement may range from just a few millimetres to tens of kilometres. Both the Highland Boundary Fault and the Southern Upland Fault are major features, giving evidence of very substantial vertical and lateral movements in their history. The lateral movements along these faults could be many tens or hundreds of kilometres while their vertical movements were also substantial. Thus estimates for the Southern Upland Fault are for vertical displacements of at least 1.5 km, and up to 5 km at its eastern end.

These two faults divide Scotland into three parts, the Highlands, the Midland Valley and the Southern Uplands, each with a quite different geological evolution. The Highland rocks, changed (or metamorphosed) from their original nature by intense heat and pressure, consist of several 'terranes' or areas with different histories. The Midland Valley can be described as a rift valley between the Highland Boundary Fault and the Southern Upland Fault. The rocks exposed in it are younger than those of the Highlands and are essentially unchanged (un-metamorphosed) from the time that they formed.

Shrinkage and disappearance of the Iapetus Ocean involved the oblique collisions of three continents called Laurentia, Avalonia and Baltica. But, so far as the Borders geology is concerned, only Laurentia and Avalonia were involved. As the continents converged, great piles of sediment, kilometres thick, that had accumulated on the ocean floor, were trapped between them. These, hardened into rocks, then underwent complex folding and fracturing in the crushing consequent on the collision, culminating in the rocks being uplifted to create a great mountain range. The processes of mountain building (orogeny) through compression and uplift of great thicknesses of sedimentary materials were first recognised in Scotland. As a result these lower Palaeozoic events are referred to as the Caledonian Orogeny. Plate-tectonic movements over the past two hundred million years have torn the resultant mountain range apart so that we now see disrupted pieces of it in northern Britain and Ireland, Norway, eastern Greenland and much of the Appalachian belt from the southern-eastern states of the USA to the maritime provinces of Canada. The uplift of mountains is inevitably accompanied by the forces of erosion that act in concert to destroy them. Rocks are no sooner raised up in the atmosphere than they are attacked by wind, rain, ice and rivers with the production of sediments. Erosion breaks up the rocks and the products get conveyed in mountain streams, to be subsequently deposited when they can no longer be borne by the flowing water. Heavier fragments (boulders, cobbles and pebbles) are deposited early as the rate of flow diminishes, whilst the finer particles have a greater chance to be carried further to be dropped on or beyond the coastlines, much of it to accumulate on the offshore continental shelves.

We have no clear idea how much sediment has been removed from the Southern Uplands in this way since the close of the Silurian. It has been estimated that about 1500 m was eroded from the Highlands, but we simply do not know how high the original Southern Upland mountains were. As the mountains were reduced in height and flow-rates in the rivers diminished, there was a generalised change in the nature of the sediments that they deposited. The coarse sediments typical of the latest Silurian and earliest Devonian gave way to rather more fine-grained deposits in the Carboniferous and Permian Periods. Whereas sedimentary deposition throughout the Ordovician and much of the Silurian took place beneath the waters of Iapetus, those of the latest Silurian and the Devonian were laid down in an utterly different environment, as subaerial sediments in a continental desert environment.

The geological map of the Borders (Fig. 3.2) shows three main outcrop areas for the Lower Old Red Sandstone. The westernmost of these is an elongate NNW–SSE strip along the Lauderdale, through which the A68 runs between Oxton and Earlston. The bright brick-red colour of the fields along the valley after ploughing is a reflection of the colour of the underlying Lower Old Red Sandstones. Lauderdale is an exhumed, ancient valley. It was originally eroded during Lower Old Red Sandstone times and filled with coarse sediment, much of which has been removed by later erosion. A second and more extensive tract runs roughly north–south from just south of Dunbar to the vicinity of Polwarth (6 km SW of Duns). The third area of outcrop lies one or two kilometres to the south-west of St Abbs and Eyemouth.

Lower Old Red Sandstone times saw extensive development of volcanoes away to the north in what were to become the Midland Valley and Grampian Terranes. Rather later, eruptions took place in the Borders, and the largest and most significant of these appears to have composed a large volcano whose remains now form the Cheviot massif straddling the Scotland–England border. Another, probably somewhat older, series of eruptions gave rise to the volcanic rocks beautifully exposed in the rugged coastal cliffs near St Abbs and Eyemouth.

More tectonic disturbances followed in the mid-Devonian although these caused much gentler folding and faulting than those of the Caledonian orogeny. Prolonged erosion in mid-Devonian times was responsible for the removal of much of the Lower Old Red Sandstone products, accounting for their scattered outcrops. Later deposition of sandstones, principally by rivers, produced the Upper Old Red Sandstone strata which lie with marked discordance upon Lower Old Red sandstones near Earlston and Duns and upon Cheviot lavas in parts of Roxburghshire. Elsewhere, e.g. between Earlston and Hawick, Lower Old Red Sandstone materials were completely removed by erosion and the younger sandstones rest directly on the steeply folded early Silurian strata.

Plate movements, although concentrated several hundred kilometres to the south, were such as to have repercussions in northern England and southern Scotland. These involved some stretching (extension) of the crustal rocks and the generation of faults attended by eruption of lavas from a suite of relatively small volcanoes lying along a WSW–ENE to NW–SE trend, and reaching from the northern shores of the Solway Firth to the Borders around Kelso and Greenlaw. The lavas from these, together with the worn-down stumps of the sub-volcanic intrusive rocks, give rise to some of the more striking scenic features of the Borders, such as the Eildon Hills. Following this volcanism, parts of the Borders continued to subside and receive river-borne sediments from the now much diminished Caledonian mountains to the north-west. Thus much of the land beneath, for example, Coldstream, Duns and Berwick-upon-Tweed became blanketed by early Carboniferous fluviatile sandstones. Virtually no record was left by the ensuing 70 m.y. of the Carboniferous Period except for some small scattered outcrops of dykes intruded at the very end of the Period.

Scotland was dry land for much of the time from the Carboniferous to the present. The final stage of the geological history of the Borders came during and after the Ice Age, and the effects of this cold time and its aftermath left memorable traces on the landscape.

CHAPTER 4

Planet Earth, its drifting continents and moving plates

The composition of the Earth

Figure 4.1 shows a cut-away model of the Earth to illustrate its structure. The distance from its centre to surface is 6341 km. Of this, the core has a diameter of approximately 3486 km whereas the enveloping zone is known as the mantle. The term 'crust' refers to the external, superficial layer of low-density rocks. Whilst the crust has very variable thickness, it rarely exceeds 40 km. The mantle, which composes by far the greater part of the Earth by volume, consists principally of rocks that are hot – but not normally molten. Just as solid, crystalline ice can flow, fluid-like, in glaciers, so too can the bulk of the mantle rocks. Thus, despite its solidity, most of the mantle is inferred to be in continual motion, churning in complex convective overturns. However, the very outermost part of the mantle is mechanically strong and does not participate in the convective overturn of the deeper mantle. It is this rigid outer part, together with the crust, that constitutes the lithosphere. The crust of the continents is much thicker than that beneath the oceans, where it is commonly about 10 km thick.

Of all the igneous rocks on Earth, basalt is the commonest. Deep dredging of rock samples, together with sea-floor drilling, has revealed that the oceans are essentially floored by basaltic lavas underlain by denser rock-types. By contrast, the continents are dominated by granitic rocks, despite the fact that these may be partly veneered by sedimentary rocks or lavas. Granitic rocks are much more silica-rich than are basalts (and the rocks composing the mantle). This fundamental difference in the chemical nature of continents and oceans, now thoroughly accepted, was not generally understood until the 1950s. Recognition led to the realisation that continents stood higher

19

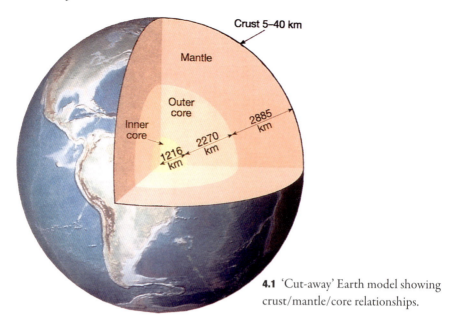

4.1 'Cut-away' Earth model showing crust/mantle/core relationships.

(and mostly emergent), because they were underlain by lighter and more buoyant materials. Granitic rocks are almost unknown in the oceans which, being floored by heavier (denser) materials, sit lower. This observation relates to the idea of isostasy: if one envisages the mantle rocks that underlie both oceanic and continental crust as acting as a fluid and that the crust 'floats' upon this substrate, it can readily be appreciated that the more buoyant continental crust will rise to higher levels than its denser oceanic counterpart. As an unloaded ship will float higher than a laden one, so do the continents 'float' higher than the rocks of the ocean floors.

In the convecting mantle the cooler, denser components descend, probably to the core–mantle boundary, whilst hotter and more buoyant parts rise to take their place. Fundamentally the convection is analogous to what one observes making porridge or marmalade! But the motion is extremely slow, fractions of a millimetre per year. If it all happens so slowly that in comparison a snail's pace would be positively supersonic, should we care? The answer is 'yes' because, slow as it is, it is the convective flow of the mantle rocks that elevates mountain ranges, rends continents and causes the opening and closing of oceans. Most geological processes proceed with unconscionable slowness, but every now and again they can be expressed with rude abruptness as in earthquakes and explosive volcanic eruptions.

The continental drift hypothesis

Before proceeding to the fate of the long-defunct Iapetus Ocean, let us first consider the oceans of the planet as they are now, and examine their relationship to the concept of plate tectonics. A glance at a globe serves as a reminder that some five-sevenths of the Earth's surface is covered by the oceans. Although we now worry about rising sea-levels, the oceans themselves appear constant and timeless so that, until a little over forty years ago, most geologists believed them to be essentially permanent fixtures on the face of the Earth. Perhaps, as was then thought, they sometimes deepened and perhaps sometimes shallowed, even to the extent of parts of the ocean floor becoming elevated and forming 'land bridges' to explain how related groups of animals and plants can be found on the opposing sides of oceans. A minority, however, argued that oceans might be created as a result of the break-up of continents and the separation of the resultant fragments. Such ideas stemmed initially from consideration of the Atlantic and its margins. Ever since the first maps of the Atlantic Ocean were drawn in the fifteenth and sixteenth centuries, explorers and geographers mused over the apparent fit of the coastlines of the Americas with those of Africa and Europe. Francis Bacon in 1620 was among those who called attention to the remarkable parallelism of the opposing coasts of the Atlantic.

The idea that continents themselves might move or 'drift' was furthered about 150 years ago by Antonio Snider who, in attempting to account for the near-identity of fossil plants in Carboniferous coal measures of Europe and North America, constructed a map in 1858 showing the Americas, Africa and Eurasia conjoined as a single continental mass. That this super-continent could have parted to allow the opening of the Atlantic Ocean appears to have been in his mind rather than the thought that some intermediate Atlantis-type lost-land had sunk in between.

Early in the twentieth century, F.B. Taylor in America and L.A. Wegener in Germany independently concluded that the continents had once been joined but then split into pieces that drifted apart. Although Taylor's ideas were published in 1910, Wegener's book *Die Enstehung der Kontinente und Ozeane* ('The Origin of Continents and Oceans'), published in 1915, made far greater impact. According to Wegener, all the continents were united some 300 Ma; subsequently Australia and Antarctica became detached, followed by incipient parting of the Americas from Africa and Eurasia some 60 Ma. By about 2 Ma an essentially modern geography had become established.

With regard to the apparent congruence of the American and Afro-European coastlines, Wegener also showed that the fit was much improved if the 200 m submarine contour ('isobath') was used rather than the coasts. It was, however, not until 1965 that a group at Cambridge University under the leadership of Edward Bullard showed that the fit became still more convincing if the 500 fathom contour (approaching 1000 m depth) was taken to mark the 'true' continental margin. A computer-generated pre-Atlantic 'fit' made by this group is shown in Figure 4.2.

4.2 An 'Atlantic fit' map showing the close match of the western and eastern sides, using the 500 fathom submarine contour. The black ornamentation indicates the degree of overlap in this computerised model. After Bullard *et al.,* 1965.

It was not merely the shape of coastlines that stimulated Wegener's ideas on continental drift, but the observation that many geological features common to the different continents could be brought into conjunction if the various continental fragments were thought of as pieces of a jigsaw puzzle that could be re-assembled.

More data on the correspondence of sedimentary and igneous phenomena, fossils, climatic evidence and fold-mountain belts on either side of the Atlantic were meticulously compiled by Alexander du Toit and published in 1937 in his book *Our Wandering Continents*. But although du Toit and Wegener won some disciples, the concept of continental drift was generally dismissed as incredible and subjected to ridicule.

Despite the superficial appeal of the hypothesis, rejection came because no realistic mechanism had been proposed whereby continents might move apart until Arthur Holmes, mentioned above for his outstanding contribution with respect to geochronology, proposed that thermal convection currents in the Earth's substratum – what we would now call the mantle – could be the responsible agency. In a paper in 1928 he postulated that hot (mantle) currents rising beneath the continents could spread out and that frictional drag would pull the continental fragments apart. His theory received widespread attention through his influential book *The Principles of Physical Geology* (1944). Nonetheless, proof of continental drift, which was to blossom into the theory of Plate Tectonics, had to wait until the 1960s when a series of discoveries in several disparate fields of enquiry converged to bring about our present state of knowledge.

Plate tectonics

In order to understand something of the story recorded by the rocks flooring the Borders, a brief résumé of plate tectonic theory is required. The lithospheric plates that compose the outermost solid shell of our planet are now believed to be in perpetual motion. There is room for confusion here: whilst the lithosphere comprises the outer shell, it is composed both of the crust and also *some* of the underlying mantle. Whilst the crust and mantle are essentially defined by their compositional differences, the lithospheric plates are defined by their rigidity rather than by their composition. In brief, the lithospheric plates can be thought of as inert, solid slabs, varying in size and thickness, that 'float' on the underlying, more mobile and 'plastic' rocks of the deeper mantle. Today, although it is now more than forty years since plate-tectonic ideas attained widespread acceptance, much detail, particularly of orogenic processes, remains to be elucidated. But the crux of the idea is that the brittle outer shell of the Earth (the lithosphere) is subdivided into about half a dozen discrete 'plates' and numerous smaller 'micro-plates' which are forever moving relative to one another. Three kinds of boundary separate the plates from each other. The first kind are the constructional boundaries along which new lithosphere is created (more or less continually) by the upwelling and crystallisation of basaltic magma along the crests of the mid-ocean ridges. Secondly there are the 'transform boundaries' along which plates slide simply past each other, i.e. no new plate material is formed and no old plate is destroyed along them. Transform boundaries include faults such as the celebrated San Andreas Fault of California. The plates do not, however, slide readily past one another; rather the plates typically lock against each other as the stresses build up. Eventually the stresses become irresistible and the plates jerk past each other, releasing pent up strain energy as earthquakes.

If there were only constructional and transform plate boundaries the planet would continuously increase in size. That it doesn't is because of the third category of plate boundaries involving the phenomenon of subduction in which the oceanic lithosphere sinks back into the mantle from which it had been created. Subduction, which will be dealt with in some more detail below, is a process of critical importance when we come to consider the origin of the Southern Uplands (Chapter 5).

Mid-ocean ridges

Information on the nature of the ocean floor has been accruing rapidly since the time of Captain Cook's great eighteenth century voyages in the Pacific. Ocean charts improved steadily, not least since echo-sounders found widespread use. Their deployment in the Pacific during the Second World War resulted in new insights into the nature of ocean-floor topography. Later, in the 1950s, marine surveying by Marie Tharp and Bruce Heezen on board the oceanographic surveying ship *Vema* led to the discovery of a remarkable submarine mountain-chain down the middle of the Atlantic. Being equidistant from the opposing coasts it was called the Mid-Atlantic Ridge (Fig. 4.3). Comparable 'ridges' (although not necessarily centrally located) were subsequently identified in other oceans, such as the East Pacific Rise and the Carlsberg and Mid-Indian Ridges in the Indian Ocean.

These 'ridges' (or 'rises'), several hundred kilometres across, link to form a circumglobal chain with a total length of ~80,000 km, rising to heights of 3 km or more from oceanic (abyssal) plains some 4 km deep. The pattern of this submarine mountain chain has been likened to that of the seam on a baseball.

Characteristic of these ridges or rises is the fissure or rift commonly found along their crests that marks the locus of recent volcanism. Many thousands of dormant or active volcanoes are located along them. Basaltic magma arising beneath these axial rifts ascends through thin vertical channels (dykes) towards the surface. Some of it erupts as submarine lava, although the bulk loses heat and solidifies before reaching the sea floor. However, in either case, the oceanic floor has new rock added to it incrementally. The solidified dykes are typically only a few metres thick, but by their steady aggregation the ridges, and hence the ocean floor in general, widens at rates generally between 20 and 100 mm per year. In this manner the ocean floor spreads symmetrically outwards on either side from the ridges, rather in the manner of paired

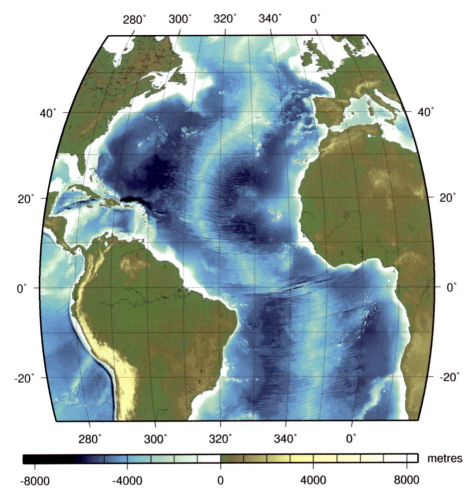

4.3 Image of bathymetry of the North Atlantic based on the satellite sea-surface altimetry (Andersen and Knudsen, 2008) model DNSC08. The symmetrical disposition of the mid-Atlantic ridge relative to the bounding continents is well exhibited. In the northern sector of the ridge the axial rift may be just discernible. www.space.dtu.dk/upload/institutter/space/data_og_modeller/bathymetry/dnsc08/at.pdf

conveyor belts. The oceanic ridges thus act as spreading 'centres' and are often called constructional plate boundaries (Fig. 4.4).

As the newly-formed oceanic basaltic crust cools through a temperature (the 'Curie Point') between 600 and 500°C, the direction and dip of the Earth's magnetic field becomes recorded within its crystalline fabric. Because the magnetic field reverses, commonly about every 200,000 to 300,000 years, when the north and south magnetic poles switch over, sea-floor spreading away from the ridge axes results in alternating strips or bands of 'normal' and 'reversed' magnetism with the contrasting

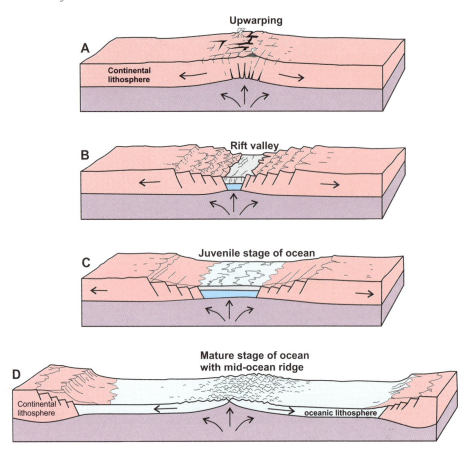

4.4 Block diagram explaining how continental pull-apart can give birth to oceans. **A.** Ascent of hot mantle rock beneath continental lithosphere causes uplift, fracturing and stretching. **B.** A rift valley develops, with magma ascending as dykes to feed volcanoes along its axis. **C.** Further extension effects complete continental rupture and generation of an embryonic ocean. **D.** The extending ocean attains maturity.

strips running parallel to the ridge axis. The magnetic strips form a pattern that is now well established and dated. Rather as tree-rings record encrypted data bearing on the life-history of the tree and increase in age from perimeter to centre, so the magnetic signatures imprinted within the ocean floor serially increase in age away from the ridge axes. Consequently the youngest parts of the ocean floor are those most recently formed at the ridge axis while the oldest parts are those furthest away.

How does an ocean start?

There is now no doubt that the mantle, from the base of the lithosphere at depths of a few tens to a few hundred kilometres down to the boundary with the core, is continuously on the move and that this motion commonly involves the ascent of more buoyant hot rocks and the descent of colder denser rocks. Whilst the precise patterns of convection remain controversial, there is agreement among both geochemists and geophysicists that convective flow is a reality and exerts drag on the lithosphere. The consequent stresses on the base of the continent-bearing lithosphere cause it to stretch. This stretching involves faulting and also makes it thinner. As the continental lithosphere thins, the underlying convecting mantle rises to shallower and shallower depths. Eventually the thinned continental lithosphere yields and the broken pieces move apart, new oceanic lithosphere is created along the parting and a new ocean can be said to have been born (Fig. 4.4)! The Red Sea provides an example of a youthful ocean created in this fashion.

Although new magma has to arise to continuously supply the expanding oceanic lithosphere, there is generally no excessive volume of basaltic magma erupted. Oceans beginning in this fashion are said to have non-volcanic or magma-poor rifted margins. To a large extent it appears that the margins bounding the newly-formed Iapetus were of this sort. In some cases, however, extension of the continental lithosphere (with concomitant pressure reduction) on unusually hot underlying mantle rocks causes the latter to start melting before continental separation commences. The basaltic magma so produced rises towards the surface, predominantly as vertical sheets (dykes) along fractures generated in the hard, brittle lithosphere. Continued thinning with its concomitant magmatism can lead to a stage where a dense parallel swarm of dykes is introduced into the crust, and erupting as very fluid lavas at the surface. The ultimate condition is reached when the dykes become so concentrated that one dyke rises within or alongside its predecessor, resulting in a situation where there is nothing but dykes cutting dykes cutting dykes! This condition defines the passage from a thinned (attenuated) continental to genuine oceanic lithosphere. Continuation of the process along an incipient 'mid-ocean ridge' then ensures lithospheric growth by spreading of ocean floor in the manner outlined in the previous section. This evolutionary sequence of events is shown diagrammatically in Figure 4.4. The Afar Depression in northern Ethiopia affords a fine example of this process in action. Here the continental crust is riven by faults and fissures brought about by extension and there is abundant evidence for recent basaltic volcanism. The Iapetus Ocean, which will be dealt with in more detail in the next chapter, is deduced to have come about through stretching and breaking apart of a great continent known as Rodinia, approximately 616 Ma. The geological evidence in the Borders, however, all concerns the later history and the closure of the ocean rather than its early opening stages.

Oceanic trenches and subduction

As explained above, the ocean floors do not go on expanding indefinitely through the spreading process, and a critical part of the plate-tectonic process is the consumption and recycling of old oceanic lithosphere. As the oceanic lithosphere migrates further and further from the spreading axis it becomes increasingly cooler and denser. Ultimately, its density can no longer be supported by the underlying mantle and it begins to sink. While this may sometimes take place spontaneously, it is currently thought that commonly some sort of tectonic 'push' is required to trigger subduction.

Deep trenches, like those peripheral to the Pacific (Fig. 4.5), develop where the oceanic lithosphere takes the plunge to be recycled, hundreds of kilometres down in the depths of the mantle. The trenches mark the sites of destructive plate boundaries. The hinge-zone, at which the oceanic plate changes from being essentially horizontal to dipping downwards into the subduction trench, is a mobile one capable of migrating back from the trench towards the mid-ocean ridge. This phenomenon, referred to as 'roll-back', appears to be common.

The trenches can be hundreds to thousands of kilometres long but measure only a few tens of kilometres in width (Fig. 4.5). Whilst the bulk of the ocean floors lie at depths of 3 to 4 km, the trenches descend to depths that can exceed 10 km (as in the Marianas Trench in the western Pacific). The Pacific trenches include those adjacent to the Chile and Peru coasts as well as those lying on the oceanic side of arcuate chains of volcanic islands ('island arcs') in the north and west of the ocean. These are typically convex toward the ocean (Fig. 4.5). The neighbouring volcanoes on their inner (concave) sides constitute the so-called 'ring of fire' around the Pacific.

Not only are the trenches associated with neighbouring volcanoes, but they are marked by vigorous seismic activity. Two seismologists who independently investigated the earthquake phenomena were Kiyoo Wadati of Japan and Hugo Benioff of the USA. They showed that the earthquake foci lie at steadily increasing depths from the outer margin of the trench inwards, defining a planar zone that generally dips away from the open ocean and beneath the volcanic arc at 40° to 60°. Although these Wadati-Benioff Zones were recognised long before the 1960s, their true nature was only established during the plate tectonic revolution of the 1960s–70s.

The crescentic or arcuate trend exhibited by most trenches is due to the fact that subduction occurs along planar surfaces that are generally oblique to the Earth's surface. Take a spherical object – for this purpose let it be an orange – and make an oblique cut in the peel. Despite the knife-cut being planar, the line of the incision will be arcuate on the surface of the orange. Only if the cut is vertical will the trace be straight!

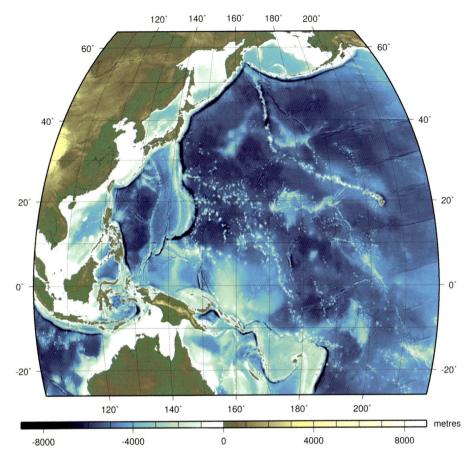

4.5 Satellite image of bathymetry of the Western Pacific (Andersen and Knudsen, 2008, model DNSC08). The intense blue-black colour marks the deep trenches where subduction is occurring. The Java Trench is seen bottom left. To the north are the Philippine, Marianas, Japan, Kurile and Aleutian Trenches. In the bottom, centre-right part of the map the most prominent feature is the NNE–SSW-trending Kermadec–Tonga Trench. (In Fig. 4.3 the Puerto-Rican and Peru–Chile Trenches are also shown). A large number of intra-plate seamounts and oceanic volcanoes occur in the west Pacific. The Hawaiian chain of oceanic volcanoes and seamounts is the notable feature trending ESE–WNW across the central Pacific, linked to the (nearly) N–S Emperor Seamount chain in the north of the ocean.

Arc magmatism

As the slab sinks into the mantle, temperatures and pressures increase. These changing physical conditions cause the water, and other components dissolved in it, to be liberated from the surface of the slab and to become absorbed into the mantle rocks above it. This lowers the melting point of the rocks to such an extent that it is responsible for melt (magma) generation above the slab by the time that it has descended

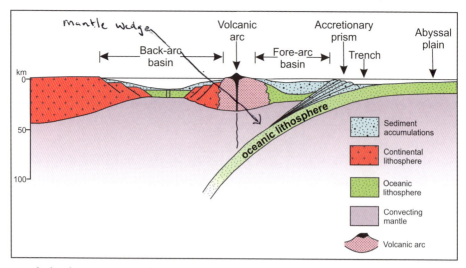

4.6 Idealised cross-section of an arc-trench system. A trench develops where oceanic litho-sphere (right), with a thin veneer of muds and deep-sea oozes, commences descend (subduc-tion) into the underlying mantle. Magmas generated above the subducting slab rise to form a volcanic arc. Materials eroded from the arc accumulate between the arc and the trench and contribute to the growing accretionary wedge (or prism). A back-arc basin may develop on the inner (concave) side of the arc, commonly between the arc and continental crust (far left). The mantle caught between the subducting slab and the volcanic arc is known as 'the mantle wedge'.

to depths of 90 to 150 km. Unsurprisingly the magmas so formed typically contain a significant amount of dissolved water. The buoyant magmas migrate upwards through the overlying mantle and crust, reaching the surface to feed volcanoes. These grow in arcuate chains parallel to, but typically 150 to 200 km distant from, the trenches, and are commonly spaced at intervals of some 20 to 50 km. Being situated above subducting lithospheric slabs, they are referred to as supra-subduction volcanoes. At the present day beautiful examples of such volcanic arcs are supplied by the Aleutian Islands, forming a pearl necklace-like northern boundary to the Pacific, and by the tropical islands of the Lesser Antilles in the Caribbean (Fig. 4.5). At the time of writing (March, 2009) fourteen supra-subduction volcanoes around the world are listed as being 'in eruption'.

However, the supra-subduction volcanoes need not necessarily compose a chain of islands. For example, along much of the Pacific coast of South America, a deep subduction trench is accompanied by a line of volcanoes on the adjacent continental margin, forming the high Andean peaks. In such a case the basement under the vol-canoes is generally composed of older rocks including folded and metamorphosed igneous and sedimentary rocks. A situation that can be thought of as intermediate between, for example, the Aleutian and the Andean cases is exemplified by Sumatra, the biggest of the Indonesian islands. Here too, youthful (active!) volcanoes lying to the inner (convex) side of a deep trench overlie relatively mature and distinctly older rocks of continental type.

The supra-subduction magmas are generally rather richer in silica than basaltic magmas, while a further distinction relates to their greater content of dissolved water – an inheritance from the hydrous fluids expelled from the descending oceanic slabs. The commonest type of lava erupted from subduction zone volcanoes is called andesite, the word taking its name from the Andean volcanoes.

As the magmas approach the surface the pressures on them lessen and dissolved water and other 'volatile' materials come out of solution as high-temperature gas bubbles. With progressive loss of these 'volatiles' the magma becomes increasingly stodgy or viscous and the hot gases, bursting to escape the enclosing magma, cause explosive eruptions. Consequently supra-subduction volcanoes tend to be notably spiteful and dangerous. Because of their generally violent eruptive behaviour, such volcanoes are generally associated with eruption, not of lava flows but of large volumes of fragmental material (loosely but erroneously called 'ashes') that may be carried high in eruptive clouds and dispersed widely in the atmosphere before settling back to earth.

Almost all the famously destructive volcanic eruptions have been from supra-subduction zone volcanoes. In the recent past these have included eruptions of Mount St.Helens in Oregon, El Chichón in Mexico and Pinatobu in the Philippines, while famous earlier catastrophic events include those of Krakatoa and Tambora in Indonesia. It is the extremely energetic eruptions of this type that should be borne in mind when evidence for early Palaeozoic volcanoes associated with the building of the Southern Uplands is presented in the next chapter.

Sedimentary wedges

Sands and muds, carried down by rivers from the eroding volcanoes and neighbouring lands and flushed out to sea, accumulate in the adjacent trenches which thus serve as highly effective sediment traps. Because of the general association of trenches and volcanic arcs, it follows that the sediments are commonly rich in volcanic materials and are thus compositionally distinct from those deposited by rivers, such as the Rhine or the Mississippi, which drain an essentially non-volcanic continental hinterland.

Everything upon the ocean floors is conveyed inexorably towards the subduction zones but, whilst the bulk will be consumed into the mantle, the lighter sediments tend to get scraped off. Rather than this occurring in a steady, continuous fashion, the process tends to be a discontinuous one in which discrete packages or wedges of

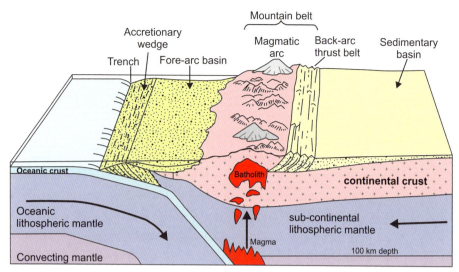

4.7 Block diagram of an idealised subduction zone with a magmatic (volcanic) arc and sedimentary basins in front of and behind it (fore-arc and back-arc basins respectively). An oceanic plate (left) is being drawn down into a subduction zone with earthquake foci at different depths. Magma generated above the descending plate (slab) rises to form a supra-subduction volcanic one above and behind it.

sediment get repetitively detached and stacked up, one beneath the other. The term 'accretionary wedges' is employed to describe these accumulations, and those forming in front of (that is to the oceanward and typically convex side of) the arcs are referred to as fore-arc suites. As the wedges stack up they become progressively rotated from being originally near-horizontal into steeper and steeper attitudes. They can accumulate to such an extent as to produce shoal ridges or even to emerge above sea-level, as in the case of the island archipelago lying off the SW coast of Sumatra. Figure 4.7 depicts a model for arc–trench systems, showing the abyssal oceanic plain, trench, accretionary prism and volcanic arc. Depositional basins forming between the prism and the volcanic arc are described as fore-arc basins.

Because of the phenomenon of 'roll-back' by the hinge zone where the oceanic lithosphere starts to be subducted, the region above it (i.e. on its concave side) can be subjected to stretching. As a result, an extensional zone can develop behind the volcanic arc, giving rise to what is called a 'back-arc basin'. Modern examples of back-arc basins include the Sea of Japan, the East China Sea and the South China Sea developed between some of the 'volcanic arcs' bowed outwards towards the Pacific and the continental mass of eastern Asia to the west (cf. Fig. 4.5).

Modern arc–trench systems show many variations on a similar theme and are shown in diagrammatic sections in Figure 4.8. This point is important with respect

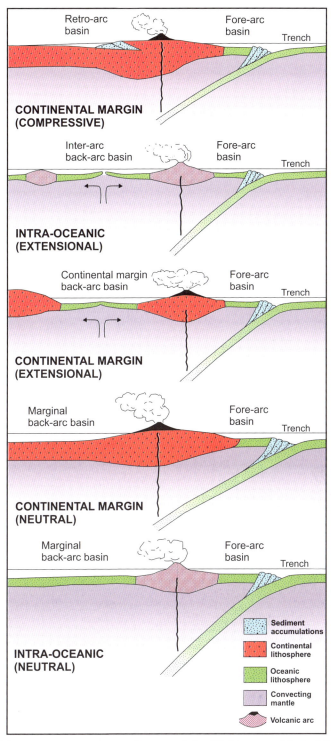

4.8 Five variations on arc-trench systems (after Dickinson and Seely, 1979).

to the Southern Uplands because, whereas there is near-universal agreement that they grew as a sequential pile of sediments largely derived from one or more volcanic arcs, this leaves almost interminable scope for argument as to whether the sediments accumulated in fore-arc or back-arc basins and whether or not the prisms ever broached sea-level to form land!

Oceanic volcanoes

So far we have described volcanism taking place along the constructive (mid-ocean) plate boundaries and the 'island arc volcanism' associated with destructive plate boundaries. Other volcanoes, however, are also common on the ocean floors well away from tectonic plate boundaries. Whilst the majority of these 'within-plate volcanoes' form 'seamounts' that remain submerged, others continue to grow and eventually emerge as islands. Some may develop over profound, localised but persistent upwellings of hot mantle rock, so-called 'mantle plumes'. The volcanoes of the Hawaiian archipelago provide well-known examples of this phenomenon (see Fig. 4.5). But countless others may denote more ephemeral thermal anomalies in the underlying mantle. Such oceanic 'within-plate volcanoes' almost certainly, as we shall see, had their counterparts in the Iapetus Ocean.

CHAPTER 5

The Southern Uplands and what they are made of

The 'bedrock' beneath the fields, forests and moors of the Southern Uplands is very largely composed of sedimentary strata. These rocks, the same described over 200 years ago by James Hutton as 'schistus', are now known to have formed from marine sediments, laid down during the Lower Palaeozoic Era within the Ordovician and Silurian Periods, between 480 and 420 Ma. As Hutton recognised, the sediments must originally have been deposited in roughly horizontal layers, with each layer younger than the one beneath; and that subsequently they were buckled and rotated so that they now stand steeply, and often vertically (Fig. 5.1).

Whilst these Lower Palaeozoic rocks are superbly exposed along the rugged coastlines (Fig. 5.2) inland exposures are sparse. Accordingly, the geologist is largely dependent on quarries, river, road and rail cuttings as well as the occasional hillside scar. In the Borders, the Grey Mare's Tail and Dob's Linn provide valuable inland sections.

Most of the sediment that was carried by a myriad of streams and rivers to the sea was mud and dirty sand. The mud changed over the millennia to shales and mud-stones whilst the sands turned into dark grey sandstones. The composition of muds and sands can be, and usually is, horrendously complex, depending on the nature of the parental rocks that were eroded to provide them. Professor H.C. Sorby of Sheffield recognised the problem a hundred years ago when he wrote: 'Possibly many may think that the deposition and consolidation of fine-grained mud must be a very simple matter, and the results of little interest. However, it is soon found to be so complex a question that one might feel inclined to abandon the enquiry, were it not that so much of the history of our rocks appears to be written in this language.' Clearly there is more to mud than meets the eye!

5.1 Folded and faulted Silurian strata. Kilburn quarry north of the Eskdalemuir-Lockerbie road (B723).

5.2 Folded and faulted Silurian strata at Pettico Wick, west of St Abb's Head, Berwickshire.

The grains that make up much of the sandstones tend to be poorly rounded or 'sharp' in builders' terminology, and also to be poorly sorted so far as their size and composition are concerned. These are the characteristics of sands that have not been carried from great distances and which, by geological standards, have accumulated rapidly. They can be described as 'immature sandstones' in distinction to those in which the sand grains have had more extended histories of transport and abrasion. In such, many of the mineral components (principally feldspars but also pyroxenes) react with water to form new flaky hydrated products, principally micas, clays and chlorite minerals, which tend to be winnowed away and carried further, ultimately to be deposited in fine-grained muds. In extreme cases after extensive alteration and winnowing of the finest products, one is left with little but grains composed of quartz, a particularly hardy or stable mineral which further sedimentary processes can only reduce to smaller grains. Rocks composed of these are called ortho-quartzites. Thus quartz tends to be a survivor in weathering and transportation. Consequently sandstones derived from weathering processes characteristic of the 'passive' continental margins are commonly composed, almost exclusively, of quartz grains, albeit cemented together by various minerals (including secondary quartz!) during subsequent solidification of the loose sands into solid rock. Thus much of the city of Edinburgh is built out of buff-coloured quartz sandstones in which the cementing material includes calcite and some iron oxides. The red sandstones, common in Glasgow, are Permian-age desert sands, again primarily of quartz grains cemented by the red iron oxide, haematite.

Such mature sandstones are a far cry from the dirty immature sandstones to which the name 'greywacke' (adapted from the German, 'grauwacke') is applied. These are of prime importance in the Southern Uplands, together with finer-grained shales dominantly composed of clay mineral particles. The particle size in greywackes is consequently very variable, and some greywacke beds contain rock fragments ranging up to several millimetres across (Fig. 5.3). The grains making up the sandstones that dominate the early Palaeozoic countryside of the Borders were transported and deposited, largely undecayed, with such rapidity, that they comprised a wide variety of other minerals besides quartz. They can be described as muddy sandstones. Greywackes make up the great bulk of the Southern Uplands, and these sombre grey to black rocks provide the common building stone for the older houses in, for example, Galashiels and Selkirk as well as for the drystone dykes that divide the fields.

In addition to grains of 'sand size', the greywackes commonly contain sharp composite gritty bits, large enough to be made up of several mineral species. These, technically referred to as 'lithic clasts', are of especial importance since they provide actual samples (albeit tiny) of the rock-types from which the sediment was derived.

5.3 Photo-micrograph of a thin-section (*c.*0.03 mm thick) of a coarse Ordovician greywacke. It is the so-called 'haggis-rock' from a quarry close to Leadburn on the A703. The slide measures *c.*50 × 20 mm and reveals a wide variety of components. Larger grains (>5 mm) include a range of sedimentary rocks (quartzites and other sandstone types, siltstones and shales). Volcanic fragments (largely andesites) are abundant and there are also volcanically-derived broken feldspar and pyroxene crystals. The slide also includes a small fragment of serpentinite.

Very coarse-grained greywackes grade into the category of conglomerates, made up of macroscopic pebbles. However, conglomerates are vanishingly rare in the Lower Palaeozoic strata of the Borders – although further south-west in Ayrshire they play an important role in the historical interpretation.

To the casual observer greywackes are unprepossessing dark grey to black rocks, scarcely worth a second glance. But then one whisky can look very like another. They are most easily examined in the drystone dykes and the walls of old buildings in the Borders. To the connoisseur, however, the subtle differences are all-important. As explained, greywackes are thoroughly immature sandstones: the sand-grains and small rock chips of which they are made up have travelled no great distances from the rocky outcrops from which they were derived. Tens of kilometres, perhaps one or two hundred, but not thousands!

Detailed analysis of the differing grains in a greywacke sample can provide a plethora of information on the nature of the rocks from which they came. Over the past thirty years a great deal of information has been compiled from painstaking forensic studies. Although the greywackes of the Southern Uplands show a broad similarity (hence the expression 'the interminable greywackes' of jaded nineteenth century investigators), there are distinct changes from one sedimentary 'package' to the next. The spectrum of rock-types that contributes to their make-up is broad. It includes a range of lavas, from basalts and andesites through to more silica-rich varieties like dacite and rhyolite, all clearly derived from volcanic terrains, most especially of the sort now characteristic of emergent island arcs and continental volcanic arcs, as exemplified by the Andes of South America. Coarse-grained granitic igneous rocks can be present, indicating that their source areas have been eroded deeply enough to reveal such rocks in outcrop. The occasional presence of serpentinite fragments is of particular interest. Serpentinite is a hydrated product of peridotite, the principal component of the upper mantle. To get serpentinite fragments in a greywacke implies that somewhere in the source area, mantle peridotites had been forced up to the surface (a process typical of uplift by faults in areas experiencing high pressures). During these vicissitudes the peridotites undergo pervasive hydration, altering to serpentinite. The latter must then be exposed at the surface to be weathered and transported in order that they may end up as fragments in a submarine greywacke deposit.

There are also various sedimentary rock fragments including shale and greywacke, reminding us that there will have been a degree of recycling and cannibalisation within the sediments. Since the processes that give rise to greywackes can be highly energetic, it is unsurprising that fast-moving submarine avalanches have the capacity to rip up fragments of previously deposited sediment.

It is to be emphasised that such sad-looking rocks as mudstones and greywackes represent repositories of a huge amount of potential palaeogeographical information. If there are 'sermons in stones', these rocks represent a vast data resource – a metaphoric Library of Congress, Library of the British Museum and the Vatican Library all rolled into one, the 'reading' and interpretation of which will occupy specialists long into the future.

There are several principal lines of research that can be utilised. The first of these is to study the nature of the bedding planes. Frequently the strata do not lie strictly parallel but exhibit furrows, grooves and embayments where water currents have eroded unconsolidated sediment prior to depositing their own cargo of detritus. Such sedimentological studies can reveal the directions that were being followed by the currents, and hence can help in indicating where the detritus may have come from. Furthermore, it can be ascertained whether the strata are 'right way' up or have been inverted by folding or faulting – bearing in mind that most of the Southern Upland strata have been tipped into steep or near-vertical attitudes!

For further study, a binocular microscope investigation of the rocks can be helpful and, even more so, the use of a petrological microscope of carefully made thin sections (usually ~0.03 mm thick) yields critical information on the nature of the crystals or rocks represented by the individual grains in the greywackes. Furthermore, high-precision quantitative data on the chemical composition of individual grains can be gathered through use of the electron micro-probe. Although the sedimentary structures reflecting their deposition are generally faithfully retained, there can be subtle changes in the mineralogy that allow estimates to be made concerning the maximum temperatures and pressures to which they were subjected during burial and folding. These show that some of the rocks reached temperatures of over 300°C, corresponding with burial to depths of up to 10 km.

Another approach is to take samples of the rocks, perhaps about 1 kg size comprising millions of sedimentary particles, crushing and subjecting them to intensive chemical analysis by the widely used technique of X-ray fluorescence spectrometry. Such bulk chemical analysis is capable of yielding rich rewards and the skilled geochemist can distinguish between sediments deposited in a variety of plate-tectonic settings. Thus submarine sediments deposited off so-called 'passive margin' environments (e.g. in the Atlantic between 'the trailing edges' of America on the one hand and Western Europe–Africa on the other) can be distinguished from those formed in tectonically active environments like those around most of the Pacific margins, in which subduction and/or continental collision is involved. It is even possible to distinguish between different subduction environments, for example between intra-ocean subduction (e.g. situations represented by the Tonga–Kermadec trench in the SW Pacific), those involving relatively mature continental island arcs (e.g. Indonesia) and those close to a fully mature continent, like those related to subduction off the west coast of South America.

Within more recent years, the Geological Survey embarked upon a study using yet another technique. The sedimentary units differ in their magnetic properties and field-mapping has been greatly facilitated by the use of magnetotelluric spectrometry. The apparatus is quite simple and can easily be carried in the field. It resembles a hand-held lantern torch and, when pressed against a rock surface and switched on, a digital signal appears on the screen that responds to the iron content of the rock. Each type of greywacke now known from its mineralogy and texture, gives an individual magnetotelluric signal, and can so be distinguished in the field. This method is highly reliable, saves time, and helped to discriminate the various tracts shown in Figure 5.4.

Apart from shales, greywackes and (scarce) conglomerates, other rock-types make a very subordinate appearance. One variety, called chert, is an extremely tough rock that commonly occurs as layers up to 10 cm thick, usually interlayered with shale.

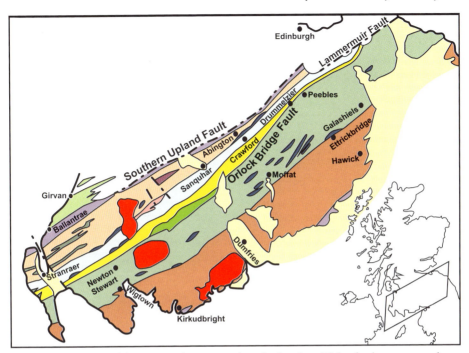

5.4 Outline geology of the Lower Palaeozoic rock in the Southern Uplands; the separate colours represent different tracts, defined by the petrography and magnetotelluric signal of the greywackes. The Northern Belt relates to the strips north of the Orlock Bridge Fault (after J. D. Floyd).

The toughness is due to the fact that chert is a relatively pure silica rock in which the minute silica crystals are intricately interlocked. The cherts formed from accumulations of marine micro-organisms called radiolarians. The skeletal remains of millions of radiolarians are required to build up a thin layer of chert. A more detailed discussion of radiolarians and these remarkable rocks is given in Chapter 7.

Volcanic rocks play a minor role in the Lower Palaeozoic formations. They are, however, of considerable importance in the unravelling of the evolutionary history of the area. Some of the basalts are, arguably, part of the oceanic lithosphere formed at the Iapetus mid-ocean ridge. Others, more probably, are products of sea-floor volcanoes. Some of the volcanic rocks are not lavas or intrusions but are 'ash-beds' composed of the altered fall-out particles from distant eruptions of emergent volcanoes. Lastly, another very quantitatively insignificant rock-type is limestone. Small but intriguing occurrences of Ordovician limestone are considered below.

Turbidites

The greywackes discussed above are defined only on the basis of their texture and composition, and not on the method of their formation. Until the early 1950s almost nothing was known about the origin of greywackes. Indeed, at one stage what are now the Southern Uplands was thought of as having originated as a gigantic area of tidal sand flats, an explanation that was generally accepted until the early 1950s. But it was not correct, and indeed the first glimmerings of a radically new understanding had begun as long ago as 1929, a year in which submarine telegraph cables off the Newfoundland coast were cut, one after the other, over a period of a few hours. Clearly, some major event must have taken place, although at the time no-one knew what had happened. But we believe we do now: such occurrences are attributed to the catastrophic destructive power of the under-sea mass-movement of unconsolidated sediments.

When rain falls on the mountains, the rivulets and streams coalesce to form serious rivers that discharge their cargoes of sediment onto the adjacent submarine shelves. As more and more sand and mud is dumped on the relatively steeply-dipping slopes leading down to the deep ocean floor, this unconsolidated material eventually becomes gravitationally unstable, peeling loose and spilling downslope as a chaotic submarine avalanche deposit. Mini-analogues of the situation can be simulated by disturbing mud on the shallows of a pond and watching the turbid material roll and swirl down into the deeper water. In the early stages the muddy cloud will involve a chaotic mixture of large and small particles but, the former will settle out first, followed by the slow separation of the finer muddy particles as the water clears. On actual continental shelves and slopes descending from volcanic arcs to deep subduction trenches, such submarine avalanches can involve millions of tons of sediment in suspension, moving briskly down to the ocean depths.

It is now evident that in 1929, and other occasions before and after, huge volumes of sediment on the outer part of the continental shelf became unstable and slid down the continental slope as a great mass of turbid sediment mixed with water, gathering momentum and breaking the telegraph cables one after the other. Eventually, as it reached the level ocean floor, the turbidity current ran out of energy, slackening its velocity and petering out, depositing the remaining sediment far from its point of origin.

Since the largest fragments settle first as coarse sands, to be consecutively overlain by finer and finer (typically clay) particles, the accumulated product shows continuous upward fining. Such submarine avalanches are called turbidity currents, and the sediments resulting from their repetition are known as turbidites. Thus a turbidite unit will be size-graded from a coarse greywacke base up through silty material to the tiny clay particles at the top that will consolidate as shale (Fig. 5.5).

The sedimentary unit resulting from a single turbidity current may be several metres thick, and such sedimentary packages may be found piled repetitively, one

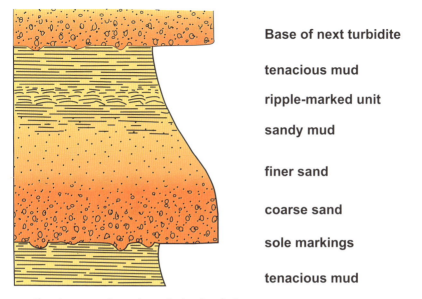

Base of next turbidite

tenacious mud

ripple-marked unit

sandy mud

finer sand

coarse sand

sole markings

tenacious mud

5.5 Sketch section through an idealised turbidite unit.

above the other, forming large submarine fans. During the passage of a single turbid-ity current, the larger, heavier particles settle first so that deposits closest to land tend to be the thickest and coarsest. They are referred to as 'proximal turbidites'. In those further from land ('distal turbidites') the particles become progressively smaller and the turbidite units correspondingly thinner, often no more than a few centimetres thick. So the sediment input onto the deep ocean floor, hundreds or even thousands of kilometres distant from the shoreline, will serially decrease with distance from the source. The turbidity currents generated off active continental margins are often trig-gered by seismic tremors.

Turbidites are abundantly represented in the Southern Uplands, and there are excellent examples to be seen in the Borders (Fig. 5.6). An example of a coarse, proximal turbidite (the so-called 'Haggis Rock', Fig. 5.3), can be seen in a quarry a few hundred metres south of Leadburn beside the Penicuik–Peebles road (A703) although a much better exposure can be seen in an old quarry at the junction of the B7007 and the B709 roads, west of Heriot, in the Moorfoot Hills, where the quarry face shows coarse greywackes composed of particles of many different kinds. (But note, the quarry face is dangerous and should only be approached with great care.)

A most instructive and easily accessible locality for examining turbidites and their associated sedimentary structures is beside the B709 road, between Innerleithen and Montbenger (NT 314 312). Here, close to the strangely named Cowpeel Bridge, is a road-cutting opened during the 1930s that still preserves as good an array

5.6 Silurian turbidite units exposed in a roadside quarry near Bentpath, on the Eskdalemuir–Langholm road (B709). The thicker (lower) parts of each unit are composed of greywacke, grading (apparently abruptly) in thin in-weathering shale units at the top.

of turbidites as can be seen anywhere. The beds, which became overturned in the Caledonian orogeny, get progressively younger from south to north (i.e. they 'young' in that direction). The relatively fine-grained greywackes here were deposited some distance from land, although the individual beds are quite thick. One should bear in mind that the thin shale beds at the top of each turbidite were originally tenacious sea-floor muds. Now, when a turbidity current spreads out over such soft sea-floor material it is, at first, energetic enough to scour and erode the latter before running out of energy and dropping its sediment load. During the erosive stage, the surface of the mud is sculpted into characteristic grooves and runnels, which are then promptly filled by greywacke sands. When the sediments are later converted into hard rock, these casts are preserved on the lower surface of the overlying greywacke and are collectively known as 'sole markings'.

If the turbidite flow was not too chaotic, the sole markings take the form of ridges aligned in the direction of the flow (Fig. 5.7a). On the other hand, where the flow is more turbulent, spinning vortices sculpt the surface of the mud into V-shaped hollows with the deepest part upstream (Fig. 5.7b). Sometimes one finds trains of these where a rapidly spinning vortex has split into several smaller ones. These infilled hollows ('flute casts') are beautifully preserved at Cowpeel Bridge on the overturned surface shown in Figure 5.8. If one imagines the bedding surface upon which these

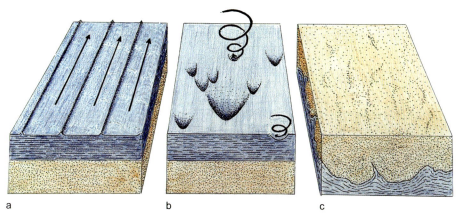

a   b   c

5.7 Mode of formation of sole markings: (a) laminar flow over tenacious mud produces parallel ridges; (b) turbulent flow with vortices over tenacious mud produces flute casts; (c) heavy sand sinks down into sloppy mud, forming load casts. Brown represents sand, blue represents mud.

5.8 Silurian turbidites. Roadside exposure at Cowpeel Bridge on the Innerleithen–Mountbenger road (B709). The exposed surface shows the underside of a turbidite unit with aligned casts of flute-marks.

sole markings are exposed rotated to its original, horizontal state, the open ends of the Vs point southwards, thus indicating that the turbidity flow came from the north. Confirmation of this is given by other sole markings within this road-side cutting. On other bedding planes at this locality there are shallower, less prominent, flute casts carved into sea-floor mud that was rather softer and more easily deformed. Also visible at some levels are bulbous, irregular masses known as 'load casts', formed where the sea-floor mud was relatively sloppy and the overlying heavy turbidite sand sank down into it (Fig. 5.7c).

Apart from sand- and mud-sized particles, turbidite currents may sometimes contain sizeable pieces of rock that have been acquired during flow. These, swept across a soft substrate, can make scratches and grooves that are referred to as 'tool marks'. Although none are to be seen at Cowpeel Bridge, an impressive, slightly curving, tool mark is clearly visible on the right-hand side of the waterfall at Dob's Linn. A remarkable instance of a near-contemporary tool mark came to light in the late 1980s during the hunt for the wreck of the German battleship *Bismarck*. The *Bismarck* was sunk in 1941 west of France and south of Ireland. As the great ship went down, she rolled over and her heavy gun turrets detached and sank through nearly 5 km to the ocean floor. At this locality the sea floor has considerable topography with a 'hill' rising ~700 m. Investigators initially found the gun turrets, but were mystified that the ship itself was not in the vicinity. In fact, the *Bismarck* had sunk in an upright position onto sloping ocean floor and then tobogganed downslope for nearly 1 km before finally coming to rest. In so doing, the hull carved out a long furrow in the oceanic sediments some 30 m wide and ~10 m deep. This extraordinary 'artificial' tool mark had survived essentially unchanged for the 48 years between the sinking and the discovery of the wreck and provides a dramatic reminder of the extreme slowness of sedimentation on the deep ocean floor.

Most of the individual turbidites at Cowpeel Bridge rest upon thin layers of shale (the tops of the previous turbidite!) and they range from less than a metre to several metres in thickness. ENKC's colleague Cecilia Taylor (Fig. 11.1), when demonstrating these rocks to palaeontology students, liked to point out that while the thin shales may have taken thousands of years to accumulate, deposition of the bulk of the overlying greywacke in the lower part of the next unit probably took no more than half an hour. Here we have a good illustration of the notion that whereas many geological phenomena take place over unimaginably long time intervals, others occur very quickly: it essentially reinforces the lessons of the *Bismarck*'s trench and the subsequent failure of sedimentation to obscure it!

Formation of submarine fans

When turbidites spill down into deep water, one after another, the successive flows may discharge into a trench, and when this is filled, will accumulate and spread as gigantic submarine fans (Fig. 5.9). While these preserve evidence of the directions

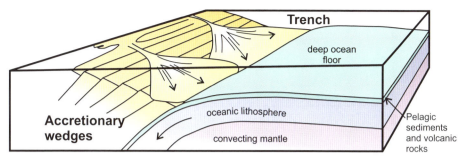

5.9 Diagrammatic representation of turbidite sediment: A) flowing into a trench and spreading as submarine fans; and B) turbiditic sediment flowing laterally along the axis of an oceanic trench. After Kelling *et al.*, 1987.

from which they come from sole markings, turbiditic sediment was often redistributed by flowing laterally along the axis of an oceanic trench. Moreover, the nature of the particles within the greywackes indicates where they come from; the submarine fans in the Southern Uplands deriving from the north have distinctive quartz-rich particles, whereas the southerly-derived turbidite fans are rich in andesitic particles from an offshore volcanic arc.

Volcanic lavas of the deep ocean

Evidence for volcanic activity on the Iapetus ocean floor is sparse and is confined to the Ordovician strata. The commonest of all magma types is basalt, a relatively magnesium-rich and silica-poor form that is erupted copiously from both oceanic and continental volcanoes. We do not know what type of volcano the lavas were erupted from, apart from the fact that they were submarine. They may have come from elongate 'fissure volcanoes' or from conical 'central type' (seamount) volcanoes. Molten basalt magma erupted under water tends to advance in pulses up to a few metres long, before the surface layer cools to form a tough restraining rind around it. These pulses advance in finger-like extensions with well-rounded fronts, tops and sides, the form of the underside being dictated by the shape of the surface over which it is flowing. When one 'finger pulse' is brought to a halt, the pressure of new magma supplied to its interior grows until the outer rind breaks and a new pulse breaks out. The rounded pillowy forms produced by this manner of advance give the general name 'pillow lava' to such sub-aqueous material. Figure 5.10 shows a diagrammatic sketch-section through lava pillows and Figure 5.11 illustrates some of the Ordovician basaltic 'pillows' (460–450 Ma) in a quarry to the south-east of the A701, between Leadburn and Romannobridge. Whilst the central parts of the

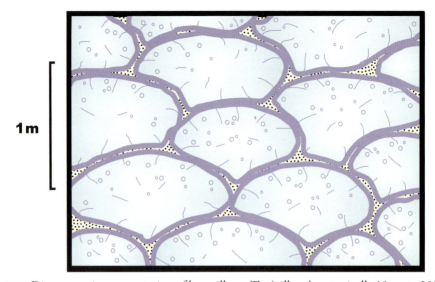

1 m

5.10 Diagrammatic representation of lava pillows. The 'pillows' are typically 10 cm to 300 cm across. As each semi-molten 'pillow' is erupted, its lower surface flows down to produce cusps between underlying (rather cooler and more rigid) pillows. Cooling of the pillows gives rise to both concentric and radial fractures. In contrast, the upper surfaces tend to have a more rounded, convex-upward form. Gas bubbles (vesicles) trapped in the cooling magma are approximately spherical. The outermost margin of each pillow becomes fast-quenched by cold water to form a glassy rind. These latter commonly shatter to a mass of small glass fragments (hyaloclastite) which, together with minerals precipitated from warm circulating waters, commonly fill the interstices between the pillows. The vesicles also become filled with minerals crystallising from these warm (hydrothermal) waters.

5.11 Ordovician pillow lavas (~455 m.y.) in quarry exposure near Noble House, SE of the A701 between Leadburn and Romanno Bridge. The rounded forms of the pillows can be discerned.

5.12 The surface of a section cut across a small 'pillow' from the exposures in [5.11]. The narrow brownish 'rind' around the pillow would initially have been shiny black basalt glass – since hydrated and then crystallised. The core would at first have been dark-grey, finely crystallised basalt containing small spherical or ovoid vesicles that later filled either with quartz (white) or chlorite (almost black). The mottled greenish-brownish colours relate to later development ('low-grade metamorphism') involving secondary growth of minerals – largely of the clay, chlorite and epidote groups.

'pillows' cool relatively slowly to give a fine-grained dark grey rock, the cooling of the outer 'rind' (usually a few millimetres thick) by the surrounding waters is so rapid that rather than crystallising, the magma supercools to a volcanic glass. Rapid contraction of the cooling glass can lead to its shattering and accumulation as masses of broken glass fragments (hyaloclastite).

The occurrences of pillow basalts and hyaloclastites in the Ordovician have, however, experienced considerable changes since they were erupted. In the crystalline lavas, various mineral species that crystallised from the molten magma are inferred to have reacted with water that permeated the rock while it was still hot. These reactions produced secondary mineral products (largely of the chlorite and epidote groups) that have greenish colours. Consequently, the pillow lavas tend to have a dullish blue-green coloration, and what would formerly have been shiny black basaltic glass composing the pillow rinds (Fig. 5.12) and hyaloclastic fragments has been degraded to a greenish assemblage involving minerals of the clay and chlorite groups. Well-bedded radiolarian cherts occur in proximity to the pillow basalts (Chapter 6).

Farther to the south-west, in Lanarkshire, still in the Ordovician but outside the area under discussion, there is an occurrence of lavas and associated fragmental volcanic rocks at Bail Hill. This is regarded as the remains of a seamount volcano that was being carried on its oceanic lithosphere conveyor towards consumption at the subduction trench.

A remarkable stratum within the Ordovician greywackes extends between 20 and 30 km along the Tweed valley. This layer, up to 35 m thick, can be followed from Glencotho, north-east through Drumelzier and Stobo to Winkston, about 3 km north of Peebles. It is extremely heterogeneous and comprises a finer-grained matrix of mudstone and volcanic tuff and rock fragments ranging from pebble size to boulders several metres long. The whole assemblage is regarded as having commenced as an unsorted slurry of rock and mud debris that cascaded down the flanks of a large seamount (or indeed, possible oceanic island). The rock and mud is inferred to have descended from shallow waters to the deep ocean floor as a submarine 'mass-flow deposit', scouring out channels in its substrate. The evidence for shallow water origin is presented below where we consider the remarkable limestone blocks in this unit.

Apart from tuff (volcanic 'ash') the unit contains masses of fine-grained volcanic rock, considered to represent lavas. These lavas, much more silica-rich than basalts, are of the type known as rhyolite. Furthermore, they have the chemical attributes of peralkaline rhyolites. Such lavas are virtually unknown among supra-subduction volcanoes but can be erupted from intra-plate volcanoes both in the 'open' oceans and back-arc basins (Chapter 3). Modern examples are known, for example from Socorro Island in the Pacific and Grand Canaria in the Atlantic. Since the best exposures are (or were!) on Wrae Hill, between the valleys of the Tweed and Holms Water about 4 km south of Broughton, they are generally referred to as the Wrae volcanics. One lava outcrop from Wrae Hill is up to 12 m thick and is overlain by up to 18 m of tuff. Rhyolitic magmas can be produced by the maturation ('fractional crystallisation') of very much larger volumes of basaltic magma. This suggests the existence of one or more large magma chambers, in turn prompting the speculation that the volcano was a large one. If so, it may have formed an oceanic island rather than merely a submarine seamount. But of the volcano itself, nothing is to be seen. Either it is somewhere buried beneath sediments or it was consumed down the presumed subduction zone. The Bail Hill seamount volcano, mentioned above, lies at much the same stratigraphic horizon in the Ordovician as the Wrae volcanic rocks. For 'modern' instances of seamounts and oceanic island volcanoes at risk of being swallowed, along with the volcanic rocks generated at the Mid-Ocean Ridges, we would point to the NW 'corner' of the Pacific where the prominent N–S chain of Emperor Seamounts awaits subduction at the junction between the Kuril and Aleutian Trenches. Christmas

Island in the eastern Indian Ocean, lying about 300 km SW of Java, is another 'at risk' oceanic volcano destined, in the fullness of time, to be swallowed down the Java Trench.

Exotic Limestones

Although limestones are present in the Ordovician of Girvan away to the south-west, they are almost entirely absent in the rocks under consideration. However, we see a rare, indeed unique, exception, in the mass-flow deposits described above associated with the Wrae volcanic rocks. As with the latter, the limestone is present as dispersed lumps or boulders within the mass-flow deposit and occur as masses up to several metres across.

The Wrae limestones, as they are usually called, have a notable place in the history of geology. Sir James Hall, a notable character in the so-called Edinburgh Enlightenment, was travelling from Edinburgh to Moffat in 1792 when he noted fossil shells in limestone being quarried at Wrae Hill. He brought these to the attention of James Hutton who was then (1795) able to state categorically that the sedimentary rocks of the Southern Uplands had formed beneath the sea. The historic Wrae Hill quarry has now largely disappeared under impenetrable conifer forest, but there is, however, a good exposure in an abandoned quarry at Glencotho. To reach it one must travel south-westwards, parallel with the Holms Water, along the minor road from Broughton to the big house at Glencotho. From here a rough track leads to the quarry on the hillside to the south. The locality is worth a visit, if only for the fine view south-westwards to the great mass of Culter Fell.

The limestones contain a special assortment of fossil species belonging to a wide range of families. These include trilobites, brachiopods, cephalopods, gastropods, bivalves and crinoids. There are also the (until recently) enigmatic little fossils called conodonts, as well as the rare graptolite specimens which help in giving more precision to the age. The whole fossil assemblage clearly denotes a relatively shallow water environment, and it has been suggested that the creatures lived in a reef biofacies that grew on the volcanic seamount or island, as suggested above. That there was close relationship between the lavas and the limestones is beyond doubt. Peach and Horne in 1899 noted limestone not only in contact with the lavas but, in one instance, actually included in lava. A shallow reef composed of shelly organisms (no corals in those days!) may well have been overrun by a lava flow.

We may further speculate that masses of reef plus lava became detached from the volcano flanks and broke up to form a mass-flow deposit, which took all of it down to the trench floor that was acting as a trap for the turbidite sedimentation. The fossils in the 'mass flow' matrix give an age of 455–450 Ma. Whereas the trilobites, up to a point, can be matched with faunas of that age from Girvan, the brachiopods are more like those found in somewhat younger carbonate mud-mounds, and there is no really clear ecological 'signal' from any of the fauna to tell us more about the environments from which they came. The limestone boulders, however, are clearly older. The tiny but very distinctive conodonts point to an age greater than 470 Ma.

Although, for the most part, fossils are vanishingly rare in the huge thicknesses of Lower Palaeozoic turbiditic sediments, their presence in the Wrae limestones gives us food for thought. Although the shallow seas surrounding the Ordovician–Silurian coastlines would have been teeming with marine life, it is unsurprising that fossils should be so rare. Transport in turbidite flows would have been highly chaotic and destructive, resulting in the elimination of the mortal remains of any organisms caught up in them. Possibly the dark greys and blacks of some of the greywackes reflect the presence of graphite – carbon derived from the breakdown of seaweed (algae) and sea-creatures.

CHAPTER 6

The birth and death of Iapetus

The idea that an ancient ocean had separated the geological units ('terranes') that now constitute Scotland and much of northern Ireland from those of England, Wales and most of Ireland was a flower that took a long time to bloom. From the late nineteenth and early twentieth centuries palaeontologists had puzzled over the contrast between the shallow-water marine fauna of the Cambrian–Ordovician as seen in Scotland (e.g. in the Girvan area and in Sutherland) and that in rocks of equivalent age in England, Wales and more southerly parts of Ireland. The former have distinctly North American affinities whilst the latter clearly possess a close relationship to those in Scandinavia and elsewhere in Europe. Some very significant factor or factors must have intervened to prevent miscegenation between these populations. Although the areas bearing the 'American-type' and 'European-type' faunas are now geographically close, the inference was that in the Cambro-Ordovician they were separated by some barrier so uncrossable so far as interbreeding was concerned that animals in each followed their own unique evolution. The explanation now accepted is that these shallow marine environments (continental shelves?) were hundreds or thousands of kilometres apart, separated by a wide ocean.

Later, when palaeomagnetism became an established science, it was realised that the magnetic signatures, encrypted into the magmatic rocks as they cooled and the sediments as they consolidated, could indicate both the latitude and the direction of the magnetic poles at the time when the magnetism became locked in. These palaeomagnetic studies helped to confirm that the two faunal domains were widely separate in the early part of the Palaeozoic.

Yet further evidence that such an ocean was not merely a flight of the imagination came from investigation of rock-suites known as 'ophiolites'. Although 'ophiolite' sounds as if it should relate to a single rock-type, this is not so. Rather the word embraces a whole clutch of associated rock-types. For many years geologists had been aware of this distinctive rock assemblage before the understanding dawned that it comprised rocks derived either from the lithosphere of fully-fledged oceans or from quasi-oceanic 'back-arc basins' (Chapter 4). The ophiolite assemblage includes submarine basalt lavas as well as serpentinites which, as explained above, are altered mantle peridotites. Normally such dense oceanic rocks would, in the fullness of time, disappear down subduction zones for recycling in the bowels of the Earth. But occasionally, during the exigencies of ocean closures, instead of going down, portions of the oceanic (*sensu lato*) lithosphere get thrust up and over continental rocks as collisions occur. Unsurprisingly in view of the brutal treatment that this entails, ophiolitic assemblages are typically seen as intensely faulted, deformed and extremely altered shadows of their former selves. They are highly relevant to the idea of an early Palaeozoic ocean, since early Ordovician ophiolites in western Norway and Newfoundland as well as in south-west Scotland and Shetland provide tangible evidence of an ocean, with or without attendant back-arc basins that have their own oceanic affinities. Thus in time all of these palaeontological, palaeomagnetic and ophiolitic studies coalesced to prove that a wide ocean had separated most of what is now North America, Greenland and Scotland from what is now most of Europe, including England, Wales and much of Ireland.

As mentioned in Chapter 4, the 1960s were years of enlightenment in which the enigmatic features of the Earth's surface (such as parallel continental margins, arcuate chains of volcanoes and their accompanying trenches, great active faults like the San Andreas, mid-ocean rises and sea-floor 'magnetic stripes') could all be comprehensively explained by the new plate-tectonic theory. In the midst of this scientific euphoria, the Canadian geophysicist Tuzo Wilson suggested that perhaps oceans had been opening and closing repetitively throughout Earth history. Following John Dewey, an English geologist, this concept became known as 'the Wilson cycle'. Dewey, in 1971, appears to have been the first to apply the plate-tectonic concept to the generation of the Caledonian mountain belt.

No doubt influenced by Dewey, Stuart McKerrow, who had been conducting diligent investigations of Silurian rocks in Britain, Ireland and Newfoundland, suggested that the Caledonian orogenesis, during which the early Palaeozoic strata had been so deformed, could have been a consequence of the closing of an ocean. This ocean was initially known as 'proto-Atlantic', but later the name Iapetus, proposed by Harland and Gayer in 1972, gained universal acceptance.

McKerrow had served aboard a Royal Navy frigate escorting North Atlantic convoys during the Second World War, making some thirty crossings of the Atlantic. In the course of this he was awarded the Distinguished Service Cross for outstanding valour. When, decades later, he was awarded a Doctor of Science degree by Oxford

University for his work on unravelling the story of Iapetus, he was proud to claim that he now had 'a DSC squared'.

The 1970s were also the years when major gas fields were discovered in the North Sea. In the course of their exploitation a gas pipe-line was to be laid across part of the Southern Uplands. Realising the import of this in terms of new geological data, McKerrow put a graduate student, Jeremy Leggett, to work on the subject. The consequence of Leggett's work was to revolutionise understanding of the structure and stratigraphy of the Southern Uplands. Two papers in the late 1970s by McKerrow, Leggett and a third collaborator, Martin Eales, presented a new plate-tectonic interpretation of the Southern Uplands. The gas-line exposures revealed a wholly new scenario from which these investigators deduced that the sediments, copiously derived from an eroding volcanic arc, accumulated in a fore-arc trench and were then scraped off in consecutive packages or slices at the subduction zone marking the northern boundary of an Ordovician–Silurian Ocean. The slices successively detached in this elegant and simple model built up what has been called an accretionary wedge(Chapter 4). Within each slice the strata get younger from bottom to top. Or rather, since each became rotated into near-verticality, we see them as getting younger from SE to NW. But, with the sequential addition of each new slice to the bottom of the accretionary wedge, the whole ensemble gets older from SE to NW! So we have the apparently contradictory situation ('the Southern Uplands paradox') that the slices themselves get younger and younger to the SE and yet, *within* each slice, the strata get younger towards the NW.

The Southern Uplands were considered to comprise a stack of at least ten of these successive slices, largely composed of turbiditic sediment dumped onto a northwestward moving ocean floor, that became sequentially jammed up behind the subduction zone rather than going down in it.

Leggett, Mitchell and McKerrow suggested that there is a modern analogue off the western coast of Burma involving just such a sedimentary stack. The oldest set of strata closest to the subduction trench is being repetitively underthrust by younger and younger sets. If sediment accumulation is rapid in situations of this sort, the whole pile can get increasingly elevated, so that what start out as deep-sea sediments are raised until a chain of islands or a ridge of dry land results. The sketch-map (Fig. 6.1) shows an island archipelago generated by this process, lying about 200 km distant from the 'volcanic arc' that forms Sumatra's mountain spine, on the convex side of the Java Trench. The archipelago includes the island of Nias: a cross-section is shown in Figure 6.2.

If there was ever a down-going Iapetus Oceanic lithosphere, a supra-subduction volcanic arc would be expected, as illustrated in Figure 4.6. The subduction trench

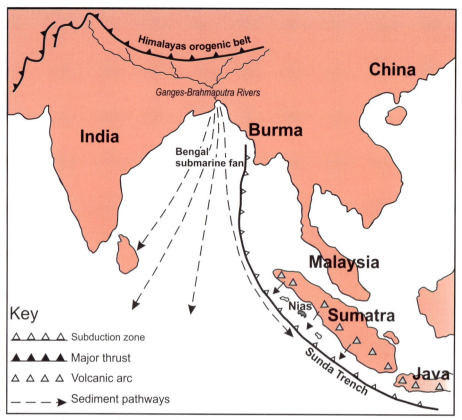

6.1. Map of Eastern Indian Ocean showing India, Sumatra and SE Asia. The subduction zone corresponds to the Sunda Trench and its northern continuation. The Nias archipelago is shown roughly midway between the trench and the volcanic arc of Sumatra and Java.

would have been south of the Southern Uplands, whilst the hypothetical volcanic arc would have lain to the north in the region now occupied by the Midland Valley. The detritus in the greywackes includes not only volcanic lavas but traces of ophi-olitic rocks, as well as of metamorphic schists suggestive of deeply eroded mountain belts. Furthermore, there are also fragments of coarser-grained intrusive rocks such as granites. The formation of granite usually implies crystallisation at considerable depth; accordingly the adjective 'plutonic' (i.e. pertaining to the realms of Pluto!) is commonly applied. However, the fragments principally fall into the category of 'hornblende granites' that can crystallise at quite shallow depths: in Indonesia, for example, some are deduced to have crystallised within a few hundred metres of the surface. Nonetheless, the 'Midland Valley' source area is generally referred to as a 'volcano–plutonic arc massif', implying that a youthful volcanic cover overlay an older and more complex foundation containing granites and metamorphic rocks.

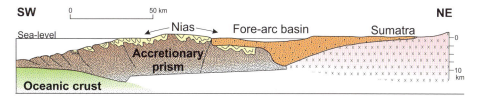

6.2 Diagrammatic section across the Nias archipelago [6.1] and the fore-arc basin between the up-raised accretionary prism and Sumatra.

It certainly appears to have been a major structure involving relatively old (possibly even Precambrian) rocks, and the bulk is considered to have developed more or less synchronously with the sedimentary basins on its southern side. Volcanoes capped this 'arc massif'. During erosion the uppermost volcanic rocks would be the first to be stripped off and, as erosion proceeds, deeper and deeper formations contribute to the sedimentary detritus. Thus in the resulting turbidite strata one should find an inverted sequence with material from the most superficial source rocks at the bottom, followed by material derived from successively deeper formations appearing in the younger strata. Indeed, sSuch a scenario can be discerned from the Southern Uplands greywackes.

The whole cycle of intrusion, uplift, erosion, transport and deposition appears to have been rapid (at least to a geologist's way of thinking). It has recently been shown that the time from crystallisation of one of the granites to its ultimate destination in a greywacke was less than 16 Ma. Thus, relative to the extended timescale of Iapetus closure, it all took place in the twinkling of an eye!

To get an idea of what such a volcano–plutonic source region may have resembled, we might think of one resembling, say, Sumatra or any of the larger Japanese islands. Such an arc massif may have shed sediments into both a back-arc basin to the north and a fore-arc basin to the south. The volcanoes of this 'massif' are presumed to have stayed active throughout the Ordovician and Silurian and indeed, the early Devonian (415–410 Ma) volcanic rocks of the Ochil and Sidlaw Hills in the Midland Valley might possibly represent prolonged continuation of activity along the same ENE–WSW zone. At this stage we should give warning for, as Leggett pointed out, there is great variety in the geometry of modern arc-trench systems (Fig. 4.8) and it would be naïve to expect to find any exact modern counterpart. And, as we shall see, the situation turns out to have been far more complex than envisaged by McKerrow, Leggett and Eales.

Whereas the Midland Valley today is some 70 km across, this geological unit (or 'terrane') is deduced to have been substantially wider in the lower Palaeozoic. Narrowing was brought about both by southerly overthrusting of the Grampian terrane from the north and northerly overthrusting by the Southern Uplands from the south. Evidence was gathered that an emergent land or island chain lay between the arc massif and the trench. In 1963 Ken Walton, an enthusiastic turbidite researcher at Edinburgh University, named this uplifted zone Cockburnland in memory of one of his recently deceased colleagues (Figs. 6.3, 6.4).

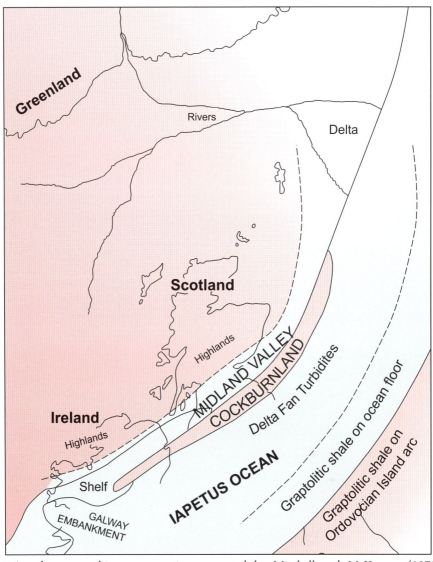

6.3 A palaeogeographic reconstruction presented by Mitchell and McKerrow (1975) showing their concept of the (hypothetical) Cockburnland resulting from an uplifted fore-arc accretionary wedge with a broad strip of accumulating turbidites to the south of it above a northward-dipping subduction zone. 'Scotland' and 'Greenland' to the north-west were part of the continent of Laurentia.

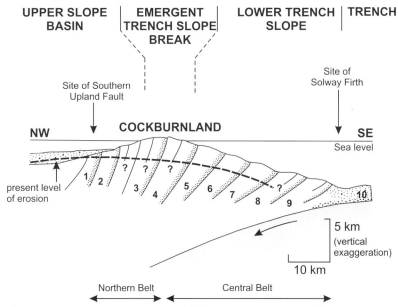

6.4 A diagrammatic section across an uplifted accretionary wedge (forming an emergent Cockburnland). The thrust slices composing the wedge (or prism) are separated by thrust faults. The dip of each increases with age (from 10 to 1) as new slices are underthrust above the subduction zone (arrowed). Dotted ornamentation at the base of each thrust-slice represents shales and cherts and, in the oldest (Ordovician) slices, basaltic lavas. The relative positions of the 'Southern Upland Fault' and 'Solway Firth' are indicated. After Leggett, 1979.

Structural features of the Southern Uplands

The Southern Uplands are defined by the Southern Upland Fault to the north and the Iapetus Suture to the south. The Southern Upland Fault is a complex structure that splits up on the northern margin of the Borders into several fractures ('splay faults'). Whereas the fault zone is now generally hidden, mainly by glacial deposits, a fine section across it was temporarily exposed during the construction of the M74, in 1990 (Fig. 6.5). Early in its history, it must have acted as a low-angle thrust fault in which Southern Upland rocks were shoved northwards, encroaching onto the Midland Valley terrane.

Many millions of years later, in the late Carboniferous to early Permian (300 ± 20 Ma), some magmatic dykes were intruded across both Ordovician and Silurian strata in the Borders. Although they involved magmas similar to basalt, these were more carbonate-rich, a characteristic that conferred on them very high fluidity and ascent rates. In their rush to the surface they ripped off fragments of their side-walls and conveyed them to shallow levels where the magmas froze to fine-grained rocks with small pieces of the acquired fragments embedded in them. Such fragments

6.5 Photo of the Southern Upland Fault taken in October 1990, during new excavations along the northbound carriageway of the M74. The main fault plane is near the standing figure on the right, with a white strip of fault gouge, and slices of various rock types are present towards the left. The fault runs westwards along the prominent valley between two hills. Photo courtesy British Geological Society.

are called xenoliths (from the Greek, meaning 'foreign stones'). They are relevant to this story because they include coarse-grained rocks composed principally of quartz and feldspar that must derive from the 'basement' beneath the Lower Palaeozoic sediments. They might come from the deeper parts of the postulated 'volcano–plutonic arc massif' of the former Midland Valley or from the Avalonian or even Laurentian continental rocks over-ridden by the Southern Uplands thrust-slices. Other xenoliths from dykes cutting the Silurian of Northern Ireland include metamorphosed volcanic rocks that have been more definitely assigned to an origin from Avalonian rocks at a depth of 20–30 km.

Later, the Southern Upland Fault evolved into a near-vertical fracture zone with the Midland Valley rocks to the north of it moving sideways, towards the SW relative to those on its southern side. Thus, to an observer south of the fault, the rocks on the far side would appear to have been moved to the left (a 'left-lateral fault'). Although each individual movement (accompanied by earthquakes) would have been limited, probably to centimetres or at most a few metres, the aggregate displacement during the course of several million years was profound. Whereas an overall shift of approximately 1500 km was proposed by McKerrow, later workers consider this exaggerated, concluding that the total movement was less than 1000 km and perhaps no more than

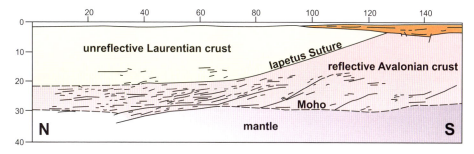

6.6 Seismic reflection profile across Iapetus Suture, showing contrast between underthrust reflective formations inferred to denote Avalonian crust in contrast to the unreflective shallower formations interpreted as representing Laurentia and overlying Lower Palaeozoic strata. The 'Mohorovičić Discontinuity' (Moho) denotes the crust-mantle boundary (after Woodcock, 2000).

300 km. In Chapter 4, the idea of transform plate boundaries was introduced, with the notorious San Andreas Fault in California as a modern example. At some stage during the late Silurian–early Devonian the Southern Upland Fault would have acted as a transform plate boundary when these lateral movements were taking place.

If the Southern Upland Fault is difficult to observe, the obscurity is worse so far as the Iapetus Suture is concerned, since in northern Britain it is coyly concealed beneath a thick cover of younger deposits. The Suture, which has been described as the most fundamental fault in the British Isles, denotes the zone along which the continental crusts of Avalonia and Laurentia became mechanically coupled. During collision the leading edge of Avalonia was partially dragged down, underthrusting the Laurentian margin. Although unseen, it can be detected by seismic and magnetic techniques, and the junction between Avalonia and overlying Laurentian crust has been shown to dip north-westwards beneath the Southern Uplands at about 45° (Fig. 6.6). Although most of the Avalonian rocks in the UK are covered by younger (mainly Upper Palaeozoic) strata, some are to be seen in the Lake District. The trace of the Iapetus Suture must roughly follow a course from the vicinity of the Cheviot massif, south-westwards to the Solway Firth. Because of the very different nature of the terranes on either side of it, the English–Scottish border coincides closely with its trace, and it essentially corresponds to the southern margin of the Borders!

Faulted subdivision of the Southern Uplands

The Southern Uplands were a cause of immense frustration to geologists throughout much of the nineteenth century owing to the relative uniformity of the rocks and the general paucity of fossils to provide relative ages and chronological sequences. Hutton recognised that the rocks he termed 'schistus' formed the hard-rock substrate and referred to the 'Schistus mountains of the South of Scotland'. After Hutton, however, no great progress was made, and for many decades it was concluded that the Southern Uplands consisted of an essentially continuous sequence of sedimentary strata, dominantly of greywackes and shales, with only sporadic occurrences of lavas, cherts and limestones to brighten up an otherwise dull scene. But by the 1850s investigators, notably Nicol and Carrick Moore, recognised that abundant tight-folding structures were involved. And, as a result of work by Sir Roderick Murchison, it became apparent that there were numerous repetitions in the sequence, either as a result of tight folding or by faulting, or both. Apart from the Wrae limestone occurrences (Chapter 4) Murchison found these rocks to be mainly unfossiliferous, other than occasional brachiopods and the remarkable fossils called graptolites.

Graptolites, which gained their name from their resemblance to pencil-drawings of mini-fretsaw blades on the bedding planes, had generally been despised, but in due course they were to provide a critical key to understanding the age sequence (or sequences) in the rocks. Graptolites, as will be shown (Chapter 9), were a group of marine organisms living in the surface waters that died out at the end of the lower Palaeozoic. Not only did they have worldwide distribution but their remarkably vigorous evolution meant that their shapes changed again and again. Consequently the graptolites at any particular horizon are unique and distinctive, making these horizons critically important global stratigraphic indicators or 'time planes'. The first to devote serious study to these fascinating organisms was Charles Lapworth, of whom more is to be told in Chapter 8. He worked extensively on the gorge section through the shales at Dob's Linn, a locality on the east side of the Galashiels–Moffat road (A708) a few kilometres south of St Mary's Loch. Lapworth's meticulous collection of graptolites through the shale sequence allowed division of the strata into zones and their attribution to the Ordovician and Silurian.

Still later in the nineteenth century further advances came about from the work, on behalf of the Geological Survey, of two other towering figures in Scottish geology, Benjamin Peach and John Horne (Fig. 6.7). These two showed that, rather than there being a continuous sequence thousands of metres thick, faulting was responsible for repetitions in the sequence. They concluded that the rocks had been compressed, concertina-fashion, into tight, steep folds, cf. Figure 5.1. Although it was clear that the Lower Palaeozoic rocks were marine and largely of deep-water origin, there was no understanding that they denoted the northern margin of a former ocean.

An important aspect of Peach and Horne's work was the recognition that the Southern Uplands were divisible into three tracts or belts, namely the Northern Belt, Central Belt and Southern Belt, each trending NE–SW (Fig. 5.4).

6.7 Dr John Horne (left) and Dr Benjamin Neeve Peach at the Inchnadmph Hotel, Assynt in September 1912.

Although in the late 1950s and early 1960s the plate-tectonic theory had yet to be formulated, it was already apparent that the Southern Uplands were sliced up by a number of major faults, first noted by Ken Walton and Gordon Craig of Edinburgh University. The surface traces of these faults also run NE–SW, parallel to the trend of the strata. The faults subdivide the lower Palaeozoic rocks of the Southern Uplands into more than thirty narrow strips or tracts (Fig. 5.4), typically 2–5 km wide, each of which tends to have its own quite distinctive idiosyncrasies. Although the faults were first identified in Scotland they, and most of the other geological features of the Southern Uplands geology, are traceable south-west into Ireland. With the new model of an accretionary prism in mind, it became apparent that these faults (now steep) would initially have been low-angled thrust faults with relatively shallow north-westerly dips. In other words, they were the structural breaks separating one set of underthrust sedimentary slices from the next. As the slices underwent progressive steepening, so also did the fault planes separating them. Because of both the faulting and the folding, a huge degree of compression must have been involved.

The fault regarded as having the most fundamental importance is that known as the Orlock Bridge Fault (Orlock Bridge itself being in Northern Ireland, Fig. 5.4).

To the north and west of it, the strata belong to the Northern Belt of the Southern Uplands, wholly composed of Ordovician rocks, whilst all those to the south and east are Silurian. Although the fault is not readily apparent to the observer, it is a zone of shearing that traverses the Borders through Drumelzier, south of Broughton and close to Peebles. Another fault (the Laurieston Fault), separating the Central and Southern Belts, runs from close to Ettrickbridge north-eastwards to just south of Galashiels. The Northern Belt (Fig. 5.4) includes a swathe of country lying south of Biggar, including Abington and Crawton as well as the hills bordering the A703 (Penicuik to Peebles), the Lammermuir Hills and part of the Moorfoots.

Left-lateral shifts along the Orlock Bridge Fault appear to have persisted through the times when the Southern Upland rocks were being scrunched up during the continental collision that terminated Iapetus. Since its movements went on to affect rocks baked by the Cairnsmore of Fleet 'granite' (in Galloway and Dumfries, away to the south-west of the Borders), intruded at 392 Ma, activity along the fault must have continued into the early Devonian.

Evidence relating to this fault derives partly from geophysical investigations and partly from geochemistry. The latter relates to some Lower Devonian granitic intrusions that penetrate the Southern Uplands. Those occurring NW of the Orlock Bridge Fault (namely the Spango, Cairnsmore of Carsphairn and Loch Doon intrusions away to the south and west of the Borders, together with the little Broadlaw intrusion in the Moorfoot Hills) have chemical compositions that contrast with those cutting the Silurian SE of the fault. These more southerly intrusions include the Cairnsmore of Fleet and Criffel intrusions away in Galloway and Dumfries as well as the very small Priestlaw and Whiteadder granites in the Borders area. (The Devonian intrusions within the Borders will be considered in more detail in Chapter 12.) It has been suggested that the Northern Belt was separated from the Central and Southern Belts by the hypothetical Cockburnland and by the Orlock Bridge Fault, and that the juxtaposition of the Northern Belt with the other two belts took place either late in the Silurian or even in the earliest Devonian, i.e. between 425 and 415 Ma. Although its true significance remains a mystery, the evidence suggests that the Orlock Bridge Fault relates to some fundamental break in the 'basement' underlying the Lower Palaeozoic strata.

The Northern Belt comprises two compositionally different types of sediment, both of mid- to late-Ordovician age. The north-westerly one contains more granitic and metamorphic components, indicative of a continental source to the north-west. By contrast, the south-eastern facies contains fragments pointing to an origin from an island arc to the south or south-east. Volcanic debris in this facies appears to derive largely from tuffaceous deposits rather than from lavas implying that volcanoes of the

6.8 Augustine volcano, a supra-subduction andesitic cone in Alaska. Some of the analogues in the late-Ordovician-early Silurian island-arcs of the closing Iapetus would also have been situated in cold and icy environments. Photograph courtesy of Jennifer Adleman, AVO/USGS.

island arc were characterised by highly explosive eruptions, much as are those of the modern 'Pacific Ring of Fire' (Fig. 6.8).

One interpretation is that sediments of the Northern Belt accumulated in a marine basin between a continental mass to the one side and a volcanic island arc to the other. This is likely to have been a 'back-arc basin' analogous to those of the present-day western Pacific (Fig. 4.5). This inferred back-arc basin was closed and deformed at the end of the Ordovician and then largely overridden by the rocks composing the Central and Southern Belts in later (Silurian) times. Although the evidence for continental and island arc sources is clear from investigations in the Northern Belt, all trace of them on the ground is now gone (Fig. 6.9). In contrast to the Northern Belt, the shales and greywackes of the Central and Southern Belts could be the erosional products of the hypothetical island arc that were dumped on its oceanward side, i.e. fore-arc deposits.

In the tectonic tracts or slices defined by the various faults, a generalised stratigraphic pattern is recognisable. A thin basal suite, composed of black shales, cherts and (in the Northern Belt) volcanic rocks, is overlain by huge volumes of rapidly accumulated turbidites. But the timing changes from NW to SE, with the turbidite influx coming in later and later in that direction. Thus in the Northern Belt the first turbidites come in during the Upper Ordovician between 450 and 440 Ma, but heading south into the Central belt, they come in progressively later. At Dob's Linn, for example, the main greywackes did not reach the area until well into the Silurian, between 440 and 430 Ma.

A group of very fine-grained sediments which accumulated across the Ordovician–Silurian time boundary constitutes the Moffat shales. These, lying principally within

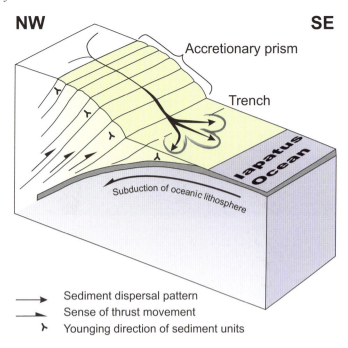

NW

SE

Accretionary prism

Trench

Iapatus Ocean

Subduction of oceanic lithosphere

→ Sediment dispersal pattern
⟶ Sense of thrust movement
⅄ Younging direction of sediment units

6.9 Accretionary prism model for the Southern Uplands (after P. Stone, 1995).

the Central Belt but also in part of the Northern Belt, have sufficient importance for us to devote a separate chapter to them (Chapter 10). They played a key role in deciphering the geological history of the Southern Uplands. The shales accumulated extremely slowly, so, although the whole group is relatively thin, its formation took many tens of millions of years. Since they were deposited in a relatively restricted area, it is possible that their depositional area was an elevated region on the ocean floor, the 'Moffat Rise' as it has been termed. If so, it was gradually encroached upon by each successive turbidite flow until it was entirely covered by them. Alternatively there may have been no such rise, but the turbidite flows reached increasingly further from their points of origin through time (Fig. 6.10).

As sediments that accumulated off the lands became gravitationally unstable, the resulting turbidite flows descended straight down the trench flanks, commonly spreading out in the depths as fan-shaped masses, as we have already noted. However, a complication is that turbidites can *also* flow lengthwise, along the axes of the trenches, sometimes for hundreds of kilometres, and can also meander very significantly. It is because of these variable factors that the field geologist, looking at a greywacke outcrop, can have severe problems trying to ascertain just where the original source area actually lay.

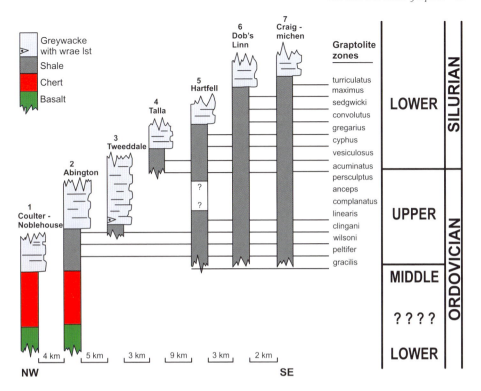

6.10 Stratigraphical sections from NW to SE, showing how the first greywackes arrive progressively later towards the centre of the depositional basin, over the Moffat Shales, calibrated against graptolite zones.

It was not until 1977 and 1978 that some workers, notably Mosely and his colleagues, began to express doubts on the widely accepted 'accretionary wedge' model for the Southern Uplands (Fig. 6.9) . Their work suggested that Iapetus closure was complete by the end of the Ordovician, thus spoiling the simple story which required subduction to have persisted up to the late Silurian. The new perspective provided a considerable stimulus to the geological community, and within the next twenty years more than a dozen researchers turned their attention to deciphering the story of the Southern Uplands and their Irish counterparts. Unfortunately, the immense amount of research that followed did not lead to any consensus. Nonetheless, all now accept that the picture *must* be considerably more complex than envisaged in the 1970s. It is worth listing the key points on which everyone *does* agree, namely that: a) the lower Palaeozoic sediments were divided up into consecutive thrust slices, ultimately to be subjected to the embrace of the tectonic 'Iron Maiden' as the continents converged; b) a subduction zone and at least one related volcanic arc were involved; and c) all lay near the northern margin of the doomed Iapetus.

Some authorities still prefer the original accretionary fore-arc wedge proposition, albeit with varying degrees of modification. Others consider that it is acceptable only so far as the Northern Belt is concerned – but not for the Central and Southern Belts. And some researchers totally reject the idea that any of the Southern Uplands strata were deposited to the south of a volcanic arc, but rather that they were formed in a back-arc basin, on the northern side of such an arc. According to this interpretation, a now vanished northern source region yielded a supply of quartz-rich sediment which cascaded down into the ocean floor forming gigantic submarine fans while, from the south, came sediment rich in andesitic material typical of island arc volcanoes. Quite often the debris from the northerly and southerly sources overlapped.

Striking evidence of contributions from a northern source, albeit to the south-west of the Borders, in Ayrshire, is given by some conglomerates containing not only pebbles but granitic boulders up to 1.5 m diameter, clearly deposited at no great distance from an eroding terrane to the NW. The granites match potential sources in Newfoundland much more closely than any Scottish granites and add to the evidence that the 'pre-Atlantic' Newfoundland sources were adjacent to the conglomerates and were subsequently shifted far to the south-west by left-lateral fault displacements.

So there is plenty of diversity in the detailed interpretations! The present authors are neither sufficiently expert nor presumptuous to judge between the validity of these various models.

Palaeogeography

At this stage let us consider the broader geographic picture. Extensional stresses exerted on the super-continent of Rodinia led to the separation of Gondwana and Laurentia at ~616 Ma and as they drifted apart, the Iapetus Ocean started to open. Although knowledge of the situation during the Precambrian and Cambrian Periods is extremely scant, an ocean probably occupied the space now taken by the Midland Valley. This appears to have been bounded to the north or west by a continental mass ('Laurentia') comprising more ancient Precambrian rocks. Representatives of these, dating from around 1800–2000 Ma, are now found in southern Greenland and the Makkovik region of Labrador.

The geography of the planet changes constantly over time, not merely because sea-levels rise and fall in response to climate change, but because the tectonic plates shift continuously in response to the inner workings of the planet. The palaeogeographic maps that can be produced have to be the result of inspired 'guesstimates' based on a host of factual data. Primarily these include the magnetic signatures which allow the palaeolatitudes to be fixed and the matches or mismatches of geological structures and rock formations of similar ages. Detailed knowledge of the space/time distribution of fossilised marine organisms allows insights into whether seaways were, or were not, opened between different continental tracts. Consequently, such maps are the results of patient detective work by many diligent investigators over the decades. The maps advance as successive approximations: they are steadily updated and made more precise as more data can be fed in.

The thinning and fracturing of the lithosphere below the Rodinian super-continent caused it to break into a cluster of smaller continental units. Those of principal concern to us were Gondwana, Laurentia and Baltica. Laurentia was made up of what are now most of North America, Greenland, Svalbard, NW Ireland, Scotland and parts of Scandinavia. Gondwana itself was, in due course, to spawn South America, Africa, India, Australia and Antarctica.

Because, by the time the earliest rocks in the Borders were laid down, Iapetus was already in decline, we have to look further afield for any evidence relating to its opening phases. In Scotland, we see evidence for its origins in the Grampians and in the far north-west. In Argyll, a huge sequence of volcanic rocks, lavas (the Tayvallich lavas) and associated sub-surface sills, is thought to mark the onset of the break-up of Rodinia. And from Durness in Sutherland south to the Sleat Peninsula of Skye, one finds former white intertidal sands (converted to quartzite sandstones) overlain by limestones deposited on the north-western margins of the youthful ocean.

Iapetus continued to widen throughout the Cambrian and is thought to have reached its maximum width (~5000 km?) in the late Cambrian to earliest Ordovician (~480 Ma) times. Figure 6.11 shows the inferred situation at 550 Ma, some 10 m.y.

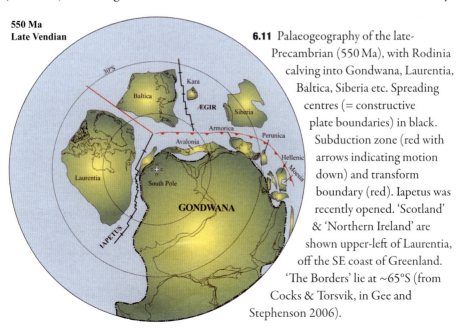

550 Ma
Late Vendian

6.11 Palaeogeography of the late-Precambrian (550 Ma), with Rodinia calving into Gondwana, Laurentia, Baltica, Siberia etc. Spreading centres (= constructive plate boundaries) in black. Subduction zone (red with arrows indicating motion down) and transform boundary (red). Iapetus was recently opened. 'Scotland' & 'Northern Ireland' are shown upper-left of Laurentia, off the SE coast of Greenland. 'The Borders' lie at ~65°S (from Cocks & Torsvik, in Gee and Stephenson 2006).

**480 Ma
Early Ordovician**

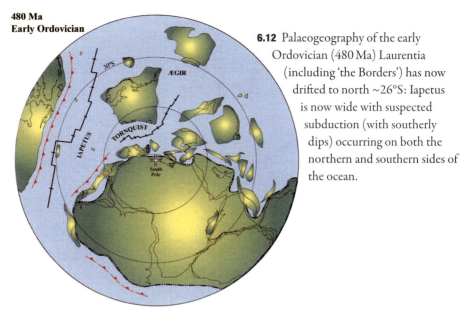

6.12 Palaeogeography of the early Ordovician (480 Ma) Laurentia (including 'the Borders') has now drifted to north ~26°S: Iapetus is now wide with suspected subduction (with southerly dips) occurring on both the northern and southern sides of the ocean.

**460 Ma
Middle Ordovician**

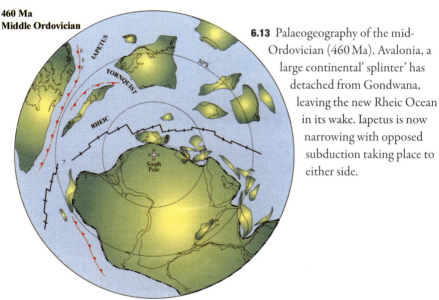

6.13 Palaeogeography of the mid-Ordovician (460 Ma). Avalonia, a large continental' splinter' has detached from Gondwana, leaving the new Rheic Ocean in its wake. Iapetus is now narrowing with opposed subduction taking place to either side.

before the start of the Cambrian Period. This, and the other palaeogeographic maps presented below, are based on reconstructions by Cocks and Torsvik (2006). In these, Scotland appears as the minuscule area to the right of southern Greenland and the Borders should be thought of as a development zone on its southern margin.

By the early Ordovician the cooler and more distal parts of the oceanic lithosphere were too dense for the underlying mantle to support and subduction commenced,

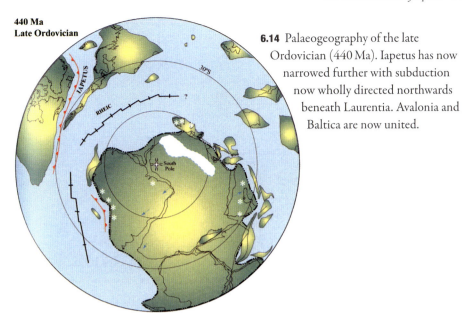

440 Ma
Late Ordovician

6.14 Palaeogeography of the late Ordovician (440 Ma). Iapetus has now narrowed further with subduction now wholly directed northwards beneath Laurentia. Avalonia and Baltica are now united.

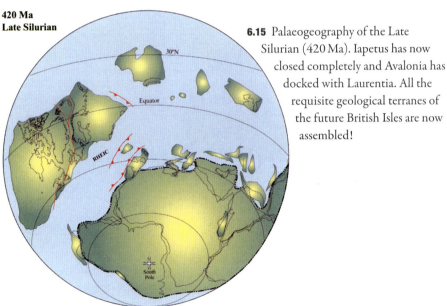

420 Ma
Late Silurian

6.15 Palaeogeography of the Late Silurian (420 Ma). Iapetus has now closed completely and Avalonia has docked with Laurentia. All the requisite geological terranes of the future British Isles are now assembled!

probably both towards the south-east under Gondwana, and north-west under Laurentia (Fig. 6.12). The rate at which oceanic lithosphere was disappearing now exceeded its rate of growth and the ocean began to shrink. Some 20 Ma later the situation is inferred to have changed as a large slice, Avalonia, became detached from Gondwana (Fig. 6.13). Avalonia included parts of what were to become north-eastern America and Canada (SE Newfoundland and a strip of country south to Cape

Cod) as well as the rest of Ireland, England and Wales, Belgium, the Netherlands and parts of northern Germany. As Avalonia converged towards Laurentia a new ocean (the Rheic) opened up behind it. Figure 6.13 depicts a portion of Laurentia in the upper left, with the elongate Avalonian continent being drawn inexorably into collision with it as the Iapetus Ocean narrowed.

Although there are spores of mid-Ordovician age, presumed to be derived from land plants, the first definite evidence for land flora comes from the Silurian. So the landscapes were generally barren, devoid of soils and subject to rapid erosion. At 460 Ma the 'Borders' lay very roughly at 40° south, a position roughly comparable to that now occupied by Tasmania or South Island, New Zealand. In the next figure (Fig. 6.14) we move forward to the Late Ordovician at around 440 Ma when there may have been only one subduction zone operating, dipping beneath Laurentia. By this stage Avalonia and Baltica had amalgamated and the future 'Borders' still lay at much the same latitude.

We now fast forward to the Late Silurian at 420 Ma (Fig. 6.15) when the combined Avalonia–Baltica continent had 'welded' itself onto Laurentia. This stage was attended by intense folding, fracturing and uplift of the materials caught between the impacting continents. The arc–trench system (or systems) associated with the westward-dipping subduction had been catastrophically crushed as the Avalonia–Laurentia vice closed!

Avalonia had been pursuing a northbound cruise, reaching sub-tropical latitudes during the Silurian when warmth-loving organisms like corals became a prominent part of the shallow marine fauna. Although shallow seas persisted, Iapetus as an ocean was now a 'dead duck' and consigned to geological history. But, as we shall see, there is evidence that its lithospheric plate continued to descend into the underworld for several tens more million years. The Scottish–Irish portions of Laurentia were now united with the English–Irish–Welsh portions of Avalonia along the Iapetus Suture and the overall geological entity that was to become Great Britain had been attained.

Concluding thoughts on the closure of Iapetus

Over the past hundred million years, what is commonly termed 'the Indian sub-continent' has been drawn into collision with central Asia through the northward-dipping subduction of the Indian Ocean lithosphere. The collision has involved a considerable degree of under-thrusting of central Asia by the 'sub-continent', a phenomenon related to the exceptional uplift of the Tibetan plateau. Further east, there is on-going convergence between Australia and SE Asia, involving an astonishing complexity of trench-arc systems (Indonesia and the Philippines) in the intervening zone. It is not an unreasonable speculation that, within a few tens of million years, Australia will have been drifted into tangential collision with SE Asia and China with concomitant tectonic mayhem. Whilst the Avalonia–Laurentia impaction may have been somewhat simpler, we may be certain that it was a lot more complicated than the original arc-trench model advocated in the 1970s by McKerrow and his collaborators, inspirational though it was at the time.

Imagine that a hedgehog is unfortunately involved in a road-traffic accident. There is no problem in identifying the mortal remains as those of a hedgehog. If, instead of a hedgehog, we consider an arc-trench system experiencing crushing injury as outlined in this chapter. The entire arc-trench system (or systems, as there is likely to have been more than one involved in the 35 million or so years in the Southern Uplands record) was hideously deformed. We cannot see the remains of the volcanic arc (or arcs) anywhere. But the myriad fragments of andesitic lavas in the greywackes provide the clearest prima facie evidence that these volcanoes *did* exist and were eroding as the sedimentary sequence accumulated. Thus we are confronted by the mystery of the lost arc – or arcs. There are two possible explanations for their absence. Either it (or they) were buried deep below slices of sedimentary rocks thrust up and over them in the climactic phase of continent collision – or it (they?) have been shunted sideways, i.e. south-westwards along one or more of the left-lateral faults (the Orlock Bridge Fault is the principal suspect!), to a position as yet unidentified. The thrust slices (accretionary wedges or not) are, however, available for our inspection, as they are the stuff of the Southern Uplands themselves. As with the unfortunate hedgehog, there is no doubt about what they were – although the question of whether they developed in back-arc or fore-arc basins or some combination of the two remains unsolved. As with our exemplary road-kill, the precise anatomical details are harder to discern. An idea of the damage inflicted can be gained from Figures 5.1 and 5.2. The fact that the closure of the continents was performed in slow motion, in millimetres per year over tens of millions of years, does not diminish its awful majesty, ultimately all driven by the force of gravity.

CHAPTER 7

The Pit and the Pendulum

Some readers may be familiar with Edgar Allan Poe's mid-nineteenth century horror story from which we have taken the title of this chapter. In this luridly haunting tale Poe describes how the hero, an officer of the Napoleonic wars, has been captured by the enemy. He finds himself lying prone, bound and immobile, and confined within a small cell, in the centre of which is a black pit. From the ceiling a pendulum with a sharp blade is sweeping backwards and forwards, all the while coming nearer. At the same time the walls of the chamber are closing in on him; the pendulum will kill him first, then the closing walls will precipitate his remains into the pit. Fortunately he is rescued just in time. Whereas we have adopted the same title we cannot promise you a scary tale. The pit to which we refer is the deep Iapetus Ocean, whereas the pendulum relates to the earliest sediments in the Southern Uplands, which show a remarkable repetitive layering of mudstone and chert, testifying to an alternation of depositional conditions as regular as the slow ticking of a grandfather clock. Before we explore these more closely, and in particular discuss why the pendulum swung in the way it did, we need to consider the order of succession of the rocks in the Southern Uplands.

Simplified stratigraphy

The succession of rocks in Southern Scotland appears relatively straightforward. The oldest rocks are basalt pillow lavas, overlain by a series of Ordovician and Silurian sedimentary rocks. At first the latter are alternating red or grey cherts and brownish mudstones, followed by dark shales, replete with fossil graptolites, which form the main subject of Chapter 9. These are overlain by a thick succession of turbidites. Such, in its simplest form, is the succession, as it has been known since the later nineteenth century, but to understand it more fully we need a detailed stratigraphical framework. Stratigraphy is concerned with the study of stratified rocks, their classification into ordered units, and their history through time. What we see in a local area, such as Southern Scotland, needs to be correlated with other areas and, in order to this the geological record has to be divided up into time periods, which should be standardised throughout the world. A primary aim of stratigraphy is to produce an accurate chronology, so that not only is the order of events known (both locally and globally), but also their dates.

One branch of stratigraphy (lithostratigraphy) concerns the order of succession of different kinds of rock in a local area, and is essential in geological mapping. The ascending sequence of rock-types in the Borders – basalt, cherts/mudstones, shales and greywackes – forms a simplified kind of lithostratigraphy. Whereas this gives essential information, it cannot give dates, nor give any basis for time-correlation with more distant areas. To do this we use the sequence of fossils found within a rock succession. This is known as biostratigraphy. At any one time in Earth history there were unique assemblages of marine organisms, chiefly invertebrates, each assemblage being characteristic for that time and no other. As time went by, these assemblages were replaced by others, through evolution and extinction, and there was thus a continuous relay of different fossils through time. In Palaeozoic sediments, brachiopods, trilobites and graptolites are commonly found; in the Mesozoic, ammonites are abundant; while in the Cenozoic snails and bivalve fossils dominate the preserved marine fauna. Some fossil species have long time ranges. The phosphatic-shelled brachiopod *Lingula,* for instance, has persisted, largely unchanged, since the Cambrian and is consequently useless for correlation. Some fossil species, however, lasted for only a fraction of geological time and the turnover rate was very high. This was true for graptolites and trilobites in the Ordovician and Silurian, and ammonites in the Mesozoic, so that they are particularly useful in this context, defining short sections of geological time known as zones. Some ammonite and graptolite zones had an average duration of less than a million years. Ideally, to be good for zonation, fossils should have a very short vertical range but a broad (intercontinental) distribution. They should also be amenable to preservation in a variety of environments and preferably, they should be independent of sea-floor environments, in other words free-swimming or floating. Some microfossils, such as conodonts and acritarchs, have great potential for stratigraphical correlation in sedimentary rocks that are poor in zonally useful macrofossils.

Both litho- and bio-stratigraphy provide a basis for organising the sequence of rocks on a global scale into defined units, so that all local as well as global events can be related to a single standard scale, and *this* is known as chronostratigraphy. In practice, a Period such as the Ordovician is divided into smaller units known as Series and these again into Stages, and Stages into Zones. In this book we need not concern ourselves with any divisions smaller than Series (other than Zones), so we simply refer to

Lower, Middle and Upper Ordovician, Silurian, Devonian and Carboniferous, which should provide an adequate framework for unravelling the history of our region.

Whereas biostratigraphy provides a relative time sequence in which it is known that fossil A always comes above Fossil B, and Fossil C always comes above Fossil B, we cannot use it to fix actual dates. But where lavas are bracketed between fossiliferous sediments, actual (or absolute) dates can be obtained radiometrically, and affixed to specific points in the fossil record. Radiometric dating and biostratigraphy are complementary, and chronostratigraphy is based on both.

The radiolarian cherts

Virtually all of the Southern Uplands rocks are of deep-water origin, and this is true of the oldest rocks, mainly Lower and Middle Ordovician, comprising basaltic lavas overlain by, or interbedded with, the cherts and mudstones already mentioned. This sedimentary sequence is rather thin (probably less than 10 m) but very distinctive. The cherts contain fossils of radiolarians (Figs. 7.1, 9.1) and conodonts (Fig. 9.3 d–f). The former are tiny unicellular planktonic organisms, with elegant perforated shells composed of silica, through which, in life, thin filaments of living material stream outwards to catch their food. Conodonts, by contrast, are made of phosphate and are the teeth, the only surviving remains of an otherwise soft-bodied, vertebrate. The abundance of radiolarians suggests comparison with the soft sediments accumulating on the abyssal plains of today's oceans, known as radiolarian oozes. We shall briefly consider different kinds of modern deep-sea oozes so that the Scottish radiolarian cherts can better be understood in context.

Deep-sea oozes

Deep-sea oozes, as their name suggests, are unconsolidated sediments, consisting mainly of the remains of micro-organisms. They were first recorded during the *Challenger* deep-ocean expedition (1872–8), and they cover about half the ocean floors. They characterise parts of the oceans that are remote and which accordingly, are largely devoid of sand and mud derived from erosion of the land since these materials have already been deposited. Consequently the oozes principally consist of the mortal remains of marine organisms that settle slowly to the ocean depths as a continuous rain of minute particles. Hence accumulation of the oozes is extremely slow. Also, as we shall consider below, they involve, albeit to a very minor but significant extent, particles from outer space.

The most widespread category is ***Globigerina ooze***, composed mainly of foraminiferans (Chapter 9) belonging to a group which lived in the upper waters of the sea, and sank to the ocean floor in uncountable numbers. Foraminiferans are protozoans

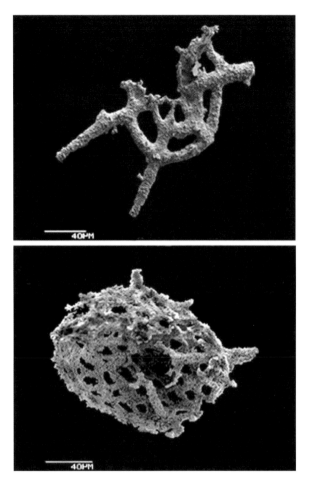

7.1 Lower Ordovician radiolarians dissolved from chert: (a) *Protoceratoikiscum clarksoni*; (b) *Inanibiguttia* species. (Photographs courtesy of Taniel Danielian, Lille University.)

with calcareous shells, often consisting of various chambers, formed one after the other. In most kinds, protoplasmic filaments stream out from the aperture of the last-formed chamber. The great majority of foraminiferans live on the sea floor, but since Cretaceous times the planktonic realm has been invaded by floating genera, of which only about sixteen live in the oceans of today, albeit in immense numbers. The commonest living form is a highly specialised foraminiferan, *Globigerina bulloides* (Fig. 9.1g). Individuals keep afloat, and also feed, using stiff protoplasmic filaments, which stream out from perforations in the five-chambered shell. Other components of *Globigerina* ooze include various foraminiferans, radiolarians, coccoliths (spherical marine algae), and pteropods (the shells of small swimming snails). Some 40% of the sea floors of the world are covered by *Globigerina* ooze, typically at depths of 2500 m to 4500 m.

A variant of this is ***Pteropod ooze***, confined to the tropics, and usually on submarine ridges at depths of less than 2000 m. Pteropods (Fig. 9.2d) have delicate shells made of calcium carbonate. The latter occurs as aragonite, which is less stable than the common form, calcite. These shells dissolve at greater depths though pressure solution, which explains their abundance in shallower waters. Other components of pteropod ooze are typical of *Globigerina* ooze. ***Diatomaceous ooze***, formed of the siliceous valves of tiny diatoms (Fig. 9.1d, f), originates from cold circumpolar waters where these minute plants swarm in great abundance.

Radiolarian ooze

Radiolarians (Figs. 7.1, 9.1h) are tiny planktonic protozoans, and they differ from foraminiferans by having siliceous, rather than calcareous tests. They are of varied and often very elegant form with perforated tests, through which thin protoplasmic filaments extrude, and with these they both feed and remain suspended in the upper waters of the sea. When they die they sink to form an ooze on the ocean floor, which entirely lacks foraminiferans or other calcareous organisms. These calcareous shells, whether originally of calcite or aragonite, have been dissolved by pressure in the deeper waters of the sea; the depth at which they do so is known as the carbonate compensation depth, 4500–5000 m below the ocean surface. ***Radiolarian ooze***, forming today, is largely confined to the eastern Pacific Ocean, forming a broad strip in the region of the equator. Living radiolarians inhabit the upper waters of the Pacific Ocean off South America. This is an area of high productivity because the nutrients which had sunk down to the sea floor are brought up again by cool currents welling up along the coast. Accordingly there is a great deal of nutrient available to stimulate phytoplanktonic productivity, the basis of all marine food chains. Radiolaria thrive in this environment, and after death most of them are swept westwards to be redeposited in waters over 4500 m deep. Radiolarian ooze thus forms at greater depths than the calcareous oozes, and is, effectively, the filtered remnants of all the small organisms that flourished in the upper waters of the sea. Although the greatest spread of radiolarian ooze is in the eastern Pacific Ocean, there are smaller areas in the Indian Ocean. Below the deepest radiolarian oozes comes *red clay*, the broken remains of radiolarians smashed by pressure and with little evidence of their original form remaining.

At what depth were the Ordovician radiolarian cherts deposited?

The radiolarians that formed the cherts in the Borders must have been exceedingly abundant. The source of the silica was very likely of volcanic origin. There is, however, an evident problem in estimating the original depth at which these cherts formed, if we try to use a direct analogy with modern radiolarian oozes. For in the Ordovician there were no planktonic foraminiferans (the earliest being Cretaceous), no diatoms (the first record is Jurassic), and few pteropods. Accordingly the cherts were not necessarily formed below the carbonate compensation depth and could have been deposited in much shallower waters. So although the field evidence suggests that they were indeed deposited in deep waters, underlain by volcanic rocks as described in Chapter 5, the precise depth remains elusive.

Are the radiolarians actually visible?

In several localities the surfaces of the cherts, when examined with a lens, or even with the naked eye, reveal little round objects. Microscopic investigation shows them to be single or double-walled spheres, some retaining radial spines. These are clearly radiolarians, but to get further detailed information it is necessary to dissolve them from the rock using 4% hydrofluoric acid. It may seem remarkable that siliceous organisms can be dissolved out of a siliceous rock, but the cherts contain a proportion of mud which facilitates the release of the radiolarians, and they actually prove to be remarkably well preserved and identifiable (Fig. 7.1). It would be very useful if they could be used stratigraphically, for most of the cherts occur only in isolated, fault-bounded slices and their time-relationships to older and younger rocks are often very uncertain. The main problem is that the Ordovician radiolarians prove to be very long-ranged forms, the same faunas being found throughout much of the early Ordovician succession. The conodonts, with which they can occur, indicate that there are certainly two levels which can be dated, one in the middle part of the Lower Ordovician, and one in the basal part of the Upper Ordovician. But as yet we do not know whether there was continuous radiolarian chert deposition throughout the time when the these sediments were being deposited, or whether there were breaks in the succession. The other stratigraphical problem is that in many instances the radiolarians have been squashed flat, and cannot be extracted for study. Accordingly, the time framework for these, the oldest sediments in southern Scotland, remains somewhat uncertain.

Sedimentary cycles in the cherts

Wherever the radiolarian cherts are visible at outcrop, they form layers a few centimetres thick, invariably alternating with layers of more shaly material, all through the sequence (Fig. 7.2). The chert layers are all very much of the same thickness, as are the shaly layers. This striking, regular alternation is also seen in Mesozoic and Cenozoic radiolarian cherts, and demands a common explanation that has been hotly debated over the years. One theory, popular for a long time, was that the layering resulted from chemical changes involving ionic segregation after sediment deposition. Yet

7.2 Radiolarian cherts showing regular alternation of chert and shale, attributed to Milankovitch cyclicity, seen in a roadside quarry near Goseland House, Coulter area south of Biggar.

original sedimentary structures, retained in both the cherts and the shales, would have been destroyed by such ionic migration, and the favoured hypothesis nowadays is that the rhythmic layering is the direct result of regular climatic cyclicity. It is now well known, especially from Cenozoic and Quaternary sediments, that warmer and cooler periods alternated. These regular climate changes are generally attributed to what is known as Milankovitch cyclicity. Three factors are involved in this. The first of these is that the Earth's orbit is not circular but slightly elliptical. Changes in this ellipticity occur in regular cycles of about 100,000 years. This means that for some part of the cycle the Earth is further away from the sun, hence cooler than at other times, when it was closer. A second kind of cycle is caused by what is known as precession of the equinoxes. Put more simply, it means that the Earth's axis wobbles like that of a spinning top. Cycles of this kind, which also affect climate, are on a 40,000 year basis. And there is a third kind of cycle, resulting from regular changes in the tilt, or obliquity, of the Earth's axis. This last kind operates over about 21,000 years. These various kinds of cycle can reinforce or suppress each other when traced through time, and these effects can be seen in deep-sea cores representing several million years.

But is the regular layering we see really the result of Milankovitch cyclicity? Some years ago a Japanese research team, working on Triassic–Jurassic chert deposits from the western Pacific, set out to test the hypothesis that rhythmic layering was truly a consequence of climatic control. Firstly they tested whether the chert layers had accumulated at the same rate as the intervening shales. Their method was ingenious. As mentioned above, there is a minor extra-terrestrial input to the sedimentation. As the Earth swings around in its solar orbit, it hoovers up a significant quantity of cosmic material. The great bulk of this is as micro-meteoritic dust that falls unnoticed to Earth. In 'normal' sediments this is totally swamped or diluted by particles coming from land sources, but on the abyssal oceanic plains, remote from land, it becomes a significant, if exotic, component. These meteoritic particles have characteristically striated surfaces which easily distinguish them from other particles, and, most importantly, they accumulate at a constant rate. It proved possible to extract them from both the cherts and the shales and to demonstrate that there are, on average, some seventy times as many cosmic dust particles in the shales as in the cherts. Since there was no way that this could have resulted from consolidation processes from soft ooze to hard rock, the cherts must have been deposited in a timespan some seventy times shorter than for the intervening shales. It seems highly probable that the cherts, with their huge numbers of radiolarians, were deposited during a relatively brief, but vast surge of productivity (the result of oceanic upwellings of nutrients?) at a particular point in a Milankovitch cycle.

Whilst it is still difficult to decide which of the three sorts of cycle was responsible, the Japanese study is at least suggestive. The stratigraphy of these strata is well constrained and it appears that the chert–shale couplets of the late Triassic accumulated in about 20,000 years, whereas the Lower Jurassic couplets, on the other hand, accumulated in about 42,000 years. In other words the Triassic sediments were picking up the second (precessional) type of cycle, while those of the Jurassic were directly influenced by the obliquity-induced, third type of cycle. As it happens, Japan was migrating northwards throughout the Mesozoic and, for as yet unknown reasons, the precessional cycle seems to have dominated at low latitudes, and the obliquity cycle at higher latitudes.

Did all three kinds of Milankovitch cyclicity operate in the Ordovician? If so, do we yet know which were recorded in the chert and shale record? So much remains uncertain that the answer remains enigmatic. But it seems safe to infer that the layering relates to Milankovitch cycles even if it is not yet possible to be more precise as to which factor or factors were involved.

CHAPTER 8

Charles Lapworth and his legacy

In the heart of the Southern Uplands, just below the highest point of the road between St Mary's Loch and Moffat, stands a small white house with a slate roof. On the outer wall is a metal plaque (Fig. 8.1) inscribed, *Birkhill Cottage, where between 1872 and 1878 Charles Lapworth recognised the value of graptolites as a clue to the geological structure of these hills. Erected by Scottish geologists, 1931.* Most travellers will pass the cottage without noticing the plaque. But for others it is a place of pilgrimage and a testament to the work of one of the most remarkable of men, not only in the development of geology, but in the whole history of science. Who, then, was Charles Lapworth and why are graptolites important in unravelling geological structures? Graptolites have already been introduced (Chapter 6) with regard to their key importance in the stratigraphic and structural interpretation of the Southern Uplands. In this chapter we shall shed more light on these matters.

Charles Lapworth (Fig. 8.2) was born in Faringdon, Berkshire, in 1842. After school years, he entered a teacher training college at Culham, near Oxford, graduating with a first-class certificate in 1864. He had read all the novels of Sir Walter Scott, many of which were set in the Borders, and, inspired to visit Scott's country, took up a teaching position in Galashiels when he was twenty-two. He held this post for eleven years but his health was not strong, and his doctor advised him to take up some outdoor pursuit, such as geology. And so he did. He began by exploring the geology of the Galashiels district in the company of his friend James Wilson, which led to his first geological publication of 1870. This earliest paper, with its simple, hand-drawn map and section clearly presages, in its meticulous presentation and its stratigraphical interest, the magnificent studies he was to undertake in the Moffat region over the next few years. His interest in the latter must have begun early, however, since he refers to the region in this first paper. Lapworth knew that his task was not going to be easy. He notes in the 1870 publication that the rocks of Southern Scotland are 'by

8.1 Plaque on the wall at Birkhill cottage, dedicated to Charles Lapworth.

Charles Lapworth, LL.D., F.R.S.

8.2 Charles Lapworth, aged about 50, from *The Geological Magazine*, 1901.

far the least known of the fossiliferous formations of this country'. He then goes on to comment that the Geological Survey maps of Scotland 'are merely indicated by a common purple tint with not the slightest attempt at division'. But as a result of his researches, and of those that followed, this was to change dramatically.

What are graptolites?

If you go to any of the localities in Southern Scotland where black shales crop out, you will see that the surfaces are often covered with rod-like, branching or coiled markings, usually with a range of fretsaw-like projections on one or both sides (Figs. 9.4–9.7). The rods or coils are known as stipes, and the projections are called thecae. The stipes themselves may take various forms; they may be single or branching, curving or spiral, and subsidiary stipes may grow out of the main one. The thecae, depending on their state of preservation, may be of various forms, short and straight, hooked or lobate, closely spaced or long and isolated from one another. The thecae in the Ordovician graptolites are usually present on both sides of the stipe and some of these forms continue, for a time, into the Silurian. Most Silurian graptolites, however, have thecae on one side of the stipe only, and these may successively change their form along the stipe. Most graptolites were originally planktonic, and only the shrubby forms known as dendroids (Fig. 9.7(1)) were moored to the sea floor. The latter were actually the ancestral type, and it was from these that the first planktonic types originated in the early Ordovician. Thereafter these fascinating creatures floated in the upper waters of the seas throughout the Ordovician, Silurian and up to the mid-Devonian, one kind succeeding another through time. They were colonial animals, with the thecae housing little feeding 'zooids' connected to each other by soft tissues ranging along the stipe. Each feeding zooid contributed to the colony as a whole. When the colonies died, they sank to the inky-black ocean depths where anoxic conditions preserved them from scavengers. The soft parts, however, decayed, and all we know of them is by analogy with some living relatives. They are, however, the only possible fossils to use for biostratigraphy in the sediments in which they occur. It is to Lapworth's eternal credit that he saw beyond the prejudices of the time and was able to develop a superlative zonal scheme based on the graptolite remains.

Until the late 1860s graptolites were not thought to be of any stratigraphical use at all. Quite a number of genera and species had been described, but often poorly and with inadequate illustrations. This, together with their frequently imperfect preservation, made it difficult to identify the species under consideration. Moreover, it was believed that they all belonged to the 'Llandeilo Series', a division of the Middle Ordovician. If so great an authority as Sir Roderick Murchison had declared this, who would have the temerity to gainsay him? Evidence of his error, however, was soon to be forthcoming. Lapworth's mentor, H. Alleyne Nicholson of Aberdeen, had written a fine memoir on graptolites in 1869 in which he showed that their strati-graphical range was much greater than had been realised, and we may suppose that he advised Lapworth to follow this up. James Hall in eastern North America had likewise begun to realise that graptolites did indeed have stratigraphical potential. Moreover, the Swedish geologist Gustav Linnarsson (1841-1881) with whom Lapworth had an increasingly vigorous correspondence, was likewise studying the distribution of graptolites in the Ordovician and Silurian of Västergötland, central Sweden. Here an unbroken sequence of Lower Palaeozoic sediments is preserved. Graptolites occur in great profusion in them, in a clear stratigraphical order. Linnarsson published his results in a series of outstanding papers from 1876 in the *Geological Magazine*, which brought them to a wide readership. Their importance was soon realised. It is recorded that Lapworth originally collected graptolites at Dob's Linn without any reference to the sequence in which they occurred. But Linnarsson advised him to collect zonally, and it is said that Lapworth set off one day in a snowstorm to test the order of succession. There is an apocryphal story that he ascended a slope at Dob's Linn (Chapter 10) putting graptolites in sequence in one voluminous pocket, and descending again, put further specimens in reverse sequence in the other pocket. He checked them when he returned home and found the sequence to be virtually identical to those established by Linnarsson and Hall. Thereafter he worked out the whole zonal sequence at Dob's Linn, and applied it to other parts of the Southern Uplands. So Lapworth came to realise that in deciphering the secrets of the Southern Uplands 'the only method found to be invariably trustworthy was that of working by means of palaeontological and lithological zones', and moreover that 'the grap-tolite was as capable as the Jurassic ammonite of being employed as a stratigraphical index'. Tragically Gustav Linnarsson, who had given Lapworth so much support and encouragement in the early days, died at the age of forty in 1881. Lapworth's dismay at the loss of his friend is evident in the articles he wrote for *Geological Magazine*. But Linnarsson's contributions through Lapworth, not only to the geology of Sweden but also to that of Britain, are inextinguishable.

Charles Lapworth was a pioneer student of graptolites and described new species of Scottish monograptids in two papers in 1876. Moreover, when the British Association held a meeting in Glasgow in 1876 he drew for the guidebook four plates illustrating all the graptolite species then known from Southern Scotland. His epoch-making paper on the geological distribution of the Rhabdophora (as he called them) contained a detailed stratigraphical analysis of all graptolites then known, from Scandinavia and eastern North America as well as Britain. But many species were poorly known or imperfectly illustrated, their descriptions being scattered in various publications, often published in obscure journals, and not easy to come by. The answer to this was to be a comprehensive monograph of all known British graptolite species, to be properly described and illustrated; but with all the pressures and commitments that he had at the time – for he had become the first Professor of Geology at what became Birmingham University – there was no way he could do it all by himself.

Unlike many of his contemporaries at the end of the nineteenth century, Lapworth actively encouraged and promoted the higher education of women. There were two, in particular,who became his closest research associates, Ethel Wood (later Dame Ethel Shakespear) and Gertrude Lilian Elles (Fig. 8.3). They were both remarkable people, and they deserve some discussion. Ethel Wood, born in 1871, studied at Newnham College, Cambridge, enrolling to read geology. There she met her life-long friend Gertrude Elles, also studying geology. Ethel became a very competent field and laboratory geologist, producing two fine papers on the graptolite-bearing Silurian rocks of the Ludlow district (1900) and the Tarannon area in Wales (1906). Around 1900 she became research assistant to Lapworth in Birmingham. From 1906 until 1914 she worked with Gertrude Elles and Lapworth on the great Monograph of British Graptolites, for which, as we shall shortly discuss, she drew all the illustrations. Then came the First World War, when she abandoned her scientific work and became intensely involved with the treatment of wounded servicemen, and pensions for the wounded and the families of those who had been killed. She co-founded the Association of War Pensions Committees and, in recognition of her tireless and effective work during and after the war, she was awarded an MBE in 1918. In 1920 she was made a Dame of the British Empire and was also awarded the Murchison Medal of the Geological Society of London. Dame Ethel never returned to science. She became a Justice of the Peace in Birmingham, and eventually she and her husband moved to a country house outside the city and settled down to farm. When hostilities broke out again in 1939 she assumed an important role on the Board of Agriculture, for which she worked actively until the end of the war. She died in 1945 at the age of seventy-four after a long life of science and devoted public service.

Gertrude Elles was born in 1872 in Wimbledon, and graduated from Newnham College, Cambridge in 1895, with First Class Honours in Geology. Her early research work was on graptolites from North Wales, the Welsh Borders, and the English Lake District. For this outstanding research she was awarded a grant from the Lyell Fund of the Geological Society of London in 1900, but could not collect it in person, since in those days women were barred from meetings. And then, under the guiding hand of Charles Lapworth, she and Ethel Wood worked for twelve years on the Monograph of British Graptolites, published by the Palaeontographical Society. During the First World War and for several years afterwards she worked actively for the British Red

8.3 *Left:* Dame Ethel Shakespear. *Right:* Dr Gertrude Elles, age about 50. (Photographs courtesy of Newnham College, Cambridge.)

Cross, and for this she was awarded an MBE. She became a Reader at Newnham College where she was renowned for her encouragement of all students and her support of women geologists in particular, as well as for the quality of her teaching. She was one of the first women to become a Fellow of the Geological Society of London in 1919, and was President of the British Association for the Advancement of Science in 1923. All her life she continued to work on graptolites and stratigraphy, especially in Wales, but she also co-operated with petrologists on unravelling the complexities of metamorphism in the Scottish Highlands. A few years before she died she wrote an extended account of her work with Ethel Wood and Lapworth, upon which we comment below. She never married and stayed in Cambridge until nearly the end of her life when, very deaf and suffering from Alzheimer's disease, she moved to Scotland where she died in 1960 at the age of eighty-eight.

We can imagine, in the halcyon days before 1914, the two women spending their summers in Birmingham, Ethel preparing her exquisite drawings with the Parkes–Lapworth microscope, specially adapted for drawing graptolites, Gertrude undertaking the descriptive work, and Lapworth supervising. Much of the time it was probably idyllic but, if Gertrude's last testament is anything to go by, Lapworth was a hard taskmaster. This testament, written in November 1957 and stored in the Birmingham University archives, was intended to form part of a study of the great man's life. Alas, Gertrude died before she could finish it. She wrote: 'While Charles Lapworth could, and did put in an enormous amount of work by day and by night when he was in the mood for it, at other times he did not seem to be able to produce anything and we just sat and talked. ...The good hours were so precious and so full of information that they would occupy the greater part of the night, until about 5 a.m., when he would suddenly stop and get Ethel to play the piano for him. Then at about 6 a.m. he would be finished for the time and would go off to rest, urging us to be ready to get back to work at about 10 a.m.' Lapworth was vastly prone to overwork for all his active life. Moreover, as Gertrude later recorded, he never fed himself properly when in the field, and all the time he was at Birkhill Cottage he lived on nothing but slices of cold porridge!

The Monograph was a magnificent work and is still in constant use today. But as Richard Fortey noted, 'The graptolites portrayed in Elles and Wood were, in truth, little more than cartoons of the original organisms. The serrated outlines of the thecae revealed little of their true structure and nothing of colony development.' They were (and are) of inestimable value in biostratigraphy even when imperfectly preserved. In the 1890s however, Gerhard Holm and Carl Wiman in Sweden discovered three-dimensional graptolites in limestones, and isolated them by dissolving them out of the rock using weak acid. Much work by Holm and his successors revealed what graptolites were like as living organisms, ushering in a new dimension for grapto-lite studies. Oliver Bulman, the Cambridge-based doyen of mid twentieth century graptolithologists, turned increasingly in this direction; he believed that the only real advances would come from the study of three-dimensional material, such as that of Koslowski in Poland, and himself in material from Girvan. Athough Lapworth, Elles and Wood must have been aware of this work, they never really took it on board, and their interests remained predominantly biostratigraphical.

Lapworth went on to do great things, though we have not the space to discuss them. He applied his graptolite stratigraphy, with great success, to the much thicker contemporaneous sequence in the Girvan district of Ayrshire, which had been deposited in nearshore conditions. He was the founder of then Ordovician System. He recognised, for the first time, the existence of colossal thrust faults in the north-west Highlands of Scotland. And when he became first Professor of Geology at Birmingham University, he investigated the geology of Shropshire and the West Midlands. Lapworth, truly a man of genius, died in 1920 at the age of seventy-eight. His influence on geology is still pervasive.

CHAPTER 9

Small fry of the surface waters

Lower Palaeozoic Plankton

At the outset, we should make two fundamental points. The first, already mentioned, is that all the sediments within our area were originally deposited on the ocean floor at depths of several kilometres. The second point is that virtually all the fossils found within the sediment were originally planktonic organisms that floated, passively drifted, or feebly swam in the upper waters of the sea. The term 'planktonic' is directly derived from the Greek πλάνκτον, which means passively drifting, and it is distinguished from nekton, which includes the strong swimmers, especially fish. After death these planktonic organisms sank slowly to the deep floor to become entombed in sediment. The reason why very little, if any, benthos (bottom-dwelling animals) lived in the stagnant Iapetus depths is that there was little, if any, oxygen available in the black ocean-floor muds. Whereas this precluded any benthos from colonising the sea floor, the planktonic remains were protected from scavenging, and the stagnant muds provided an eminently suitable environment for their preservation. The turbidite sediments that accumulated later were loose, and benthic creatures could not colonise such unstable surfaces. Since, all through this chapter, we shall be referring to planktonic organisms, ancient and modern, it seems appropriate that we should begin with a résumé of the nature of marine plankton living today. We shall discover how dramatically different were the Ordovician and Silurian plankton from that of the present day.

Marine plankton of today

The most fundamental component is the ***phytoplankton***, the tiny photosynthetic algae of various kinds, which form the basis of all the food chains in the sea. Since sunlight is absorbed very quickly in sea water, and fades with depth, living phytoplankton are confined to the upper few metres. Some typical phytoplankton types are shown in Figure 9.1. Many of these algae are naked, and have no protection. They

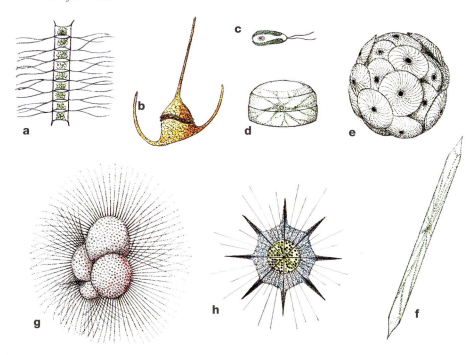

9.1 Living phytoplankton (a–e) and protozoans (f, g): (a) *Chaetoceras decipiens*, a diatom; (b) *Ceratium tripos*, a dinoflagellate; (c) *Isochrysis galbana*, a flagellate alga; (d) *Coscinodiscus*, a diatom; (e) *Cyclococcolithus leptoporus*, a coccolith; (f) *Rhizosolenia styliformis*, a diatom; (g) *Globigerina bulloides*, a foraminiferan; (h) An acantharian radiolarian. All much enlarged.

may be tiny, like the little motile *Isochrysis* which propels itself with a pair of flagella; some algae are very much larger, and others, such as *Chaetoceras*, form long chains of identical cells. Others have more robust external coverings. Dinoflagellates like *Ceratium* have stout organic walls, and are often remarkably spiny. They propel themselves with delicate whip-like appendages (flagella), one of which beats constantly in a groove round the body while the other, if present, is free. Diatoms differ in having siliceous shells. *Coscinodiscus* looks like an elaborately sculptured pill-box, but other forms such as *Rhizosolenia* have the form of an extended rod, sharp at both ends. Coccolithophorids, of which *Cyclococcolithophora* is figured here, are round, with several elaborate wheel-shaped calcareous projections, known as coccoliths.

The **zooplankton** consists of a multitude of small animals, most of which likewise inhabit the upper waters, though some live deeper. There are those creatures which are permanent inhabitants of the plankton (Fig. 9.1g, h; Fig. 9.2a–g) but there are also planktonic larval stages (Fig. 9.2h–l) of many kinds of larger animals which are benthic, in other words living on the sea floor, or nektobenthic, swimming but feeding from the sea bottom. Of these, firstly there are the protozoans (Fig. 9.1g, h) which

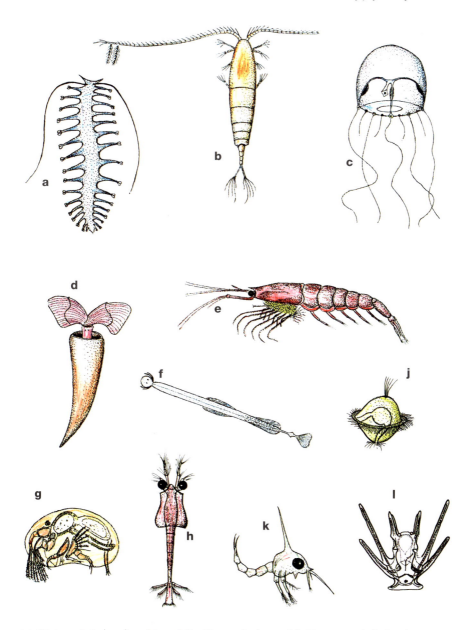

9.2 Living adult (a–g) and larval (h–k) zooplankton: (a) *Tomopteris helgolandicus*, a worm (length 10 mm); (b) *Euchirella*, a copepod (length 4mm); (c) *Thaumantius macotica*, a hydrozoan medusa (width 10 mm); (d) *Clio*, a pteropod (length 25 mm); (e) *Euphausia superba*, a peracarid crustacean (length 25mm); (f) *Sagitta elegans*, an arrow worm (length 25mm); (g) *Cypris*, a living ostracod (length 2mm); (h)) *Thysanoessa marina* a larval euphausid crustacean (length 8mm); (j) trochophore larva of a benthic worm (length 2mm); (k) zoea larva of a benthic crab (length 3mm); (l) *Echinometra lucumbria*, echinopluteus larva of an echinoid (height 0.7mm).

have single cells. These include calcareous-shelled foraminiferans (already encoun-tered in Chapter 7) such as the well known *Globigerina*. Here the jelly-like body is confined within linked chambers of calcite; the walls of these are perforated, to allow stiff rods (pseudopodia) to stream out radially; these catch smaller organisms which are then ingested. There are also the radiolarians with their elegant perforated shells made of silica; these live and feed in a similar way to the planktonic foraminiferans. Both foraminiferans and radiolarians can be preserved as fossils but, as outlined in Chapter 7, the foraminiferans, sinking to the ocean floor after death, dissolve below a critical level, the carbonate compensation depth (at about 4500 m below the surface in the Atlantic and somewhat deeper in the Pacific). The radiolarians, however, are unaffected, and may accumulate at considerable depth, forming deep-sea oozes. The Ordovician radiolarian cherts of Southern Scotland are ancient, lithified equivalents of these modern oozes.

Of the other planktonic animals, the majority are crustaceans, and they are sin-gularly diverse. These include ostracodes (Fig. 9.2g) and copepods (Fig. 9.2b), and shrimp-like forms. These latter, the peracarids, are very important, and many of these, such as the euphausiids (Fig. 9.2e) are permanent residents of the plankton. But there are also the larval forms of benthic crabs (Fig. 9.2k) and lobsters, which go through many transformations before they settle down on the sea floor. Sea-squirts (tunicates) and echinoderms are represented by their larvae, as are marine snails and sea-slugs. There are also jelly-like planktonic worms such as the transparent 'arrow-worms', *Sagitta* (Fig. 9.2f), and the elegant *Tomopteris* (Fig. 9.2a), provided with swimming paddles. A few tomopterids have been found in the Granton 'shrimp-bed' in Edinburgh, which takes the origin of this group at least as far back as the Lower Carboniferous. Large jellyfish (scyphozoans) float passively or swim by pulsation as an important component of the plankton, and there are also the large siphonophores, provided with 'swimming bells'. But there are also great numbers of little jellyfish which are hydrozoans (Fig. 9.2c). These latter exhibit an 'alternation of generations' life style. The 'hydroid' phase consists of a colony of asexual zooids, small tentaculated feeding animals, all connected to each other; the hydroids are sessile, i.e. 'fixed' to the sea floor or to seaweeds. From the lower part of the colony little jellyfish are budded off from a central column, often enclosed in a bottle-shaped sac. These escape and become the dispersal and sexual phase of the hydrozoans, producing eggs and sperm from vesicles located on the underside. This brief review includes only the dominant components of the modern zooplankton, but there are many others; we have not space to include them all.

Many planktonic animals rise to the surface at night and descend to deeper waters during the day – a vertical trip of 60 m down, and the same distance up, is not uncommon. One reason why they undertake this diurnal migration is that the zooplankton feed at night and sink during the day when they can less easily be seen. Another suggestion is that since surficial water masses may move at a different speed to deeper ones, vertically migrating animals have a better chance of a wide dispersal. This is not, however, the whole story. Nor is it fully understood why phosphorescence and light flashing are so common in both phyto- and zooplankton; it does not seem very sensible for any planktonic organisms to advertise their presence in this way. So, if so much remains to be understood about living plankton, might it not seem futile to try to understand that of the Lower Palaeozoic? But in fact quite a lot is actually known about the history of the plankton through time, and in particular that of the Ordovician and Silurian.

History of the marine plankton

The main preservable components of the Lower Palaeozoic plankton are the stick-like serrated graptolites, which we shall consider below in more detail. And since there are no planktonic graptolites living today, the composition of ancient plankton was clearly different. So in this section we shall investigate how plankton has changed through time, what components of the Lower Palaeozoic plankton may have persisted to the present day, and we shall study the continuous relay of small floating animals and plants that have led through time to the present-day planktonic communities. This may seem, at first sight, to be somewhat of a digression from the geological evolution of the Borders. But the fossil evidence that we have in southern Scotland lends itself very well to an analysis of the evolution of plankton through time, and it forms part of the overall story.

It is clear enough that the preservation potential of most living phyto- and zoo-plankton is very low. Let us first consider the phytoplankton. Undoubtedly there were small floating algae in the Lower Palaeozoic, but they seem to have been of the naked kind, and they do not preserve well. Much of the organic carbon in the black shales is undoubtedly of algal origin but has degraded and preserves no detail. The only common components of the Lower Palaeozoic phytoplankton are the acritarchs (Fig. 9.3b). These are usually spherical, with numerous projections, and they are normally considered to be some kind of algae. They had strong organic walls making them very resistant to decay. They can be extracted from the rock using acids, in specialised laboratories. The earliest known forms date from the late Precambrian, and they range up into the Cenozoic. Acritarchs are very diverse, widespread, and have short time-ranges, and they have been used stratigraphically with great success, especially in the Palaeozoic. What other components of Ordovician and Silurian phytoplankton there may have been remains unknown. But the plankton of the Lower Palaeozoic

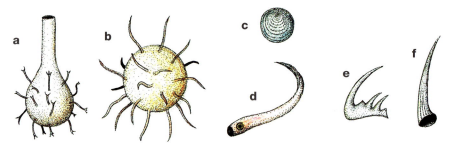

9.3 Palaeozoic planktonic organisms: (a) Silurian chitinozoan, *Gotlandichitina villosa*; (b) Silurian acritarch *Baltosphaeridium spinigerum*; (c) epiplanktonic brachiopod, Ordovician (length 4mm); (d) reconstructed conodont-bearing early vertebrate, Carboniferous (length 40mm); (e) Ordovician conodont elements *Cordylodus*; and (f) *Acontiodus*. (c, d, g much enlarged.)

contained no diatoms, dinoflagellates, or coccoliths. The earliest known diatoms are Jurassic in age; the first dinoflagellates come from the Triassic Period and the earliest coccolithophorids are Cretaceous. Much of the chalk of southern England, France and Northern Ireland is entirely composed of coccoliths; they became a vastly important component of the phytoplankton during Upper Cretaceous times.

Of the zooplankton living in the upper waters of the seas today, only radiolarians were present as common preservable components in the Ordovician and Silurian. The earliest belong to the lowermost Ordovician. Whereas the fossil record of benthic foraminiferans likewise goes back to the Ordovician, the planktonic types did not arise until the Jurassic. Undoubtedly the plankton of the Lower Palaeozoic also included various kinds of small hydrozoan jellyfish. There may have been several kinds of worms, but we have no strong evidence. The kinds of larvae that were present can only be inferred from the kinds of benthic invertebrates then living; echinoderms, snails, bivalves, and brachiopods. Swimming among the plankton were conodonts – small, active, lamprey-like creatures (probably vertebrates) of which the apatite 'teeth' are normally preserved (Fig. 9.3d–f). Pteropods (Fig. 9.2d) are a kind of marine snail, adapted for a swimming life, and these were certainly present, though rarely, in the Ordovician and Silurian. Quite commonly found in Palaeozoic sediments are chitinozoans (Fig. 9.3a), flask-shaped chitinous-shelled organisms, usually with numerous projections. Occasionally many hundreds have been found, linked together. They are generally regarded as the egg-cases of some extinct organism, and they may have been planktonic or benthic. Like the acritarchs, these have proved extremely valuable in biostratigraphy. But although there were small crustaceans (*Caryocaris*), sometimes found in Silurian black shales, there were none of the elegant 'shrimps' so common today; these did not originate until the latest Devonian. Because their shells were thin and poorly reinforced by mineral covering, their chances of preservation were slim in fully marine environments. They are sometimes found, however, in sediments of fresh or brackish water origin, such as the Lower Carboniferous 'shrimp-beds' of Gullane and Granton in the Edinburgh district, and in the Carboniferous lake sediments of the Borders (Chapter 14). In the Lower Palaeozoic the most abundant preservable components are the graptolites. In subsequent sections we shall consider the grapto-lites themselves in detail; how they were constructed, how they lived and evolved, and what bearing they have upon geology. The Scottish Borders provide almost as good a place as any in which to study them.

Graptolites as Lower Palaeozoic Plankton

Preservation and study of graptolites

Most graptolites are known from flattened specimens alone, as is typical in the Southern Uplands (Figs. 9.4, 9.5, 9.6a, b). A standard example (though actually this one does not occur in Scotland) is *Didymograptus extensus* (Fig. 9.6a). Here we see the conical sicula, the first-formed part of the colony, in the centre, and the two extended stipes, with their undersides lined with thecae. In each theca there was a small, ten-taculated zooid which fed on smaller organisms. All the zooids were linked together by a 'common canal' running within the upper part of each stipe. The colony is now known as the rhabdosome. This kind of preservation, with all its limitations, was

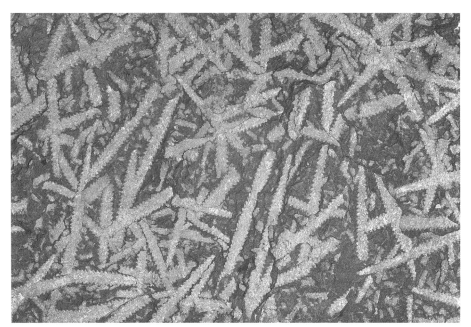

9.4 *Orthograptus truncatus*, an Upper Ordovician monospecific assemblage from the *clingani* zone at Dob's Linn ×1.5. (Photo by Bill Crighton.)

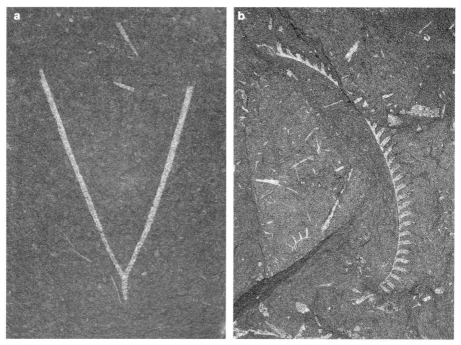

9.5 (a) *Dicranograptus ramosus,* Ordovician, *clingani* zone ×1; (b) *Rastrites peregrinus,* Lower Silurian, *triangulatus* zone. Dob's Linn ×2. (Photos by Bill Crighton.)

basically what was available to Charles Lapworth, Gertrude Elles and Ethel Wood. Their main interests were biostratigraphical, and although they recognised that graptolites were planktonic, they were not involved with either the ecology or anatomy, but only with the rapidly evolving shapes of the rhabdosomes.

As explained in Chapter 8, graptolites may be preserved three-dimensionally in a variety of ways. Not uncommonly the original hollow interior of the graptolite is filled with pyrite, and even though the original exoskeleton has disappeared, the original shape of the colony, the form of the thecae and how they link to the stipes is retained. When rock slabs containing such specimens are cracked open they often reveal to perfection the convex outer surface of the pyrite infilling, which replicates the internal surface so that it is an internal mould. The concave counterpart, however, moulds the external surface and is an external mould. If liquid rubber latex is poured into this mould and allowed to set, it can be stripped off when dry, and faithfully replicates the external surface, showing excellent details of the original structure.

Even better are instances where the actual skeleton of the graptolite, though chemically altered, is still preserved, and material of such a kind is often found in fine limestones (Fig. 9.6c-f). Individual specimens can be isolated by solution with weak acids, and then rendered partially translucent by 'clearing' with various chemical agents.

Such preparations were first made in the 1890s by the Swedish palaeontologists Wiman and Holm. The latter prepared immense amounts of Swedish material, but he died before he could complete his researches. The bulk of the descriptive and illustrative work was taken up by Oliver Bulman of Cambridge University in the 1930s, and published in a splendid series of papers. Bulman believed that the only real advances would come from the study of three-dimensional material. He then described and illustrated the only three-dimensional fauna so far known from Scotland (Laggan Burn, near Girvan). This fauna had been discovered in calcareous nodules by Charles Lapworth when he worked on the geology of the Girvan District. As Wiman and Holm had done, he broke the concretions into smaller pieces and dissolved them in strong hydrochloric acid followed by weak hydrofluoric acid, thus liberating the specimens. Finally he cleared them with concentrated nitric acid and potassium chlorate to render them translucent, and mounted them in balsam. In addition, to show the relationships between the thecae, he made closely-spaced serial sections of selected specimens, normal to the axis, and built up actual models in beeswax, layer after layer from these sections. Today, the interpretation of serial sections can be done so much more easily, and so much less malodorously, by using computers.

Meanwhile Roman Kozłowski, working in Warsaw in the 1930s, was studying equivalent three-dimensional material from the Holy Cross Mountains of central Poland. Whereas Kozłowski's monograph of 1948 contained perhaps the most detailed descriptions and illustrations ever published on graptolites, he was also able to show that the long-extinct graptolites are related to a group of living benthic colonial animals, the pterobranchs, which are distantly related to the vertebrates. This new conception revolutionised our knowledge, showing that graptolites were of a much higher grade of organisation than had been suspected. There are two genera of living pterobranchs, *Rhabdopleura* (Fig. 9.6g) and *Cephalodiscus*. They are both small sessile creatures, the former consisting of a horizontal tube from which several vertical tubes arise, the last one being closed off by a terminal bud. The skeleton is composed of thin, strap-like 'fusellar rings' as in graptolites. In each tube there is a zooid, with a pair of feathery tentacles with which it catches food. Each zooid is connected to the horizontal tube by a flexible stalk and can extend to feed in the water, or retract for protection into its own tube. Although no graptolite zooids have been preserved, it is quite likely that they resembled those of *Rhabdopleura*. *Cephalodiscus*, on the other hand, has a hemispherical spiny skeleton, with many cylindrical cavities, in each of which resides a zooid. The zooids here are quite similar to those of *Rhabdopleura*, but they are not connected to each other, and free to wander over the surface of the colony. Pterobranch colonies are made of collagen, the strong yet flexible protein of which our own tendons are constructed. In the very rare instances in which the actual material of which the graptolite skeleton (periderm) is composed is preserved, revealing its ultrastructure, it proves also to be collagen.

Structure and geological history of graptolites

We have seen that the skeletons of graptolites consist of a series of interlinked tubes with thecae bearing little polyps or zooids. Moreover, we have seen that most graptolites were planktonic. So let us consider first of all the three-dimensional structure of a standard Upper Silurian graptolite, *Saetograptus chimaera* (Fig. 9.6c). The skeleton consists of a sicula with an extended thin rod, the nema, arising vertically from the top of the sicula, and on one side of this is an array of similar-looking thecae. Each theca has two parts, the inner protheca, and the outer metatheca. There are no partitions between the successive prothecae, so that they are all linked by a common canal. The metathecae, however, are separated by walls, so that each zooid is set in its own little cup, each linked to the other by soft tissue along the common canal.

At the microscopic level, the sicula and the thecae were formed from thin rings or half-rings of collagen, collectively forming a characteristic pattern of development known as fusellar tissue. Each ring was probably secreted in a single day from the outer wall of the zooid. There is a second kind of skeletal material, however, known as cortical tissue, very thin in such a graptolite as this and external to the fusellar tissue. In graptolites where it is thicker this tissue is seen to form 'bandages', swathing the surface as in an Egyptian mummy, and at different angles. These cortical bandages can only have been applied from the outside, painted on externally. Improbable though this may seem, this was done by the zooids, tethered on long flexible stalks, but capable of creeping out of the thecae to apply the collagen in a long strip from their mouth region (Fig. 9.6h). While this may sound ludicrously improbable, a very similar

---

Opposite **9.6** Structure of graptolites (a-f) and pterobranchs (g, h): (a) *Didymograptus extensus* (L .Ord.), as preserved flattened in shale, showing a central sicula and two extended stipes, lined with thecae on the lower surface; (b) *Nemagraptus gracilis* (basal U. Ord.), flattened; (c) *Saetograptus chimaera* (U. Sil.) showing sicula and three thecae, constructed of sub-parallel bands of fusellar tissue – the small diagram shows its appearance when flattened; (d) *Climacograptus inuiti* (U. Ord), a scandent biserial graptoloid in three-dimensional preservation and with fusellar rings faintly showing; (e) part of a dendroid stipe showing the large autothecae, smaller bithecae, and the pipelike connecting tube, or stolotheca; (f) *Monograptus turriculatus* (L. Sil.) which forms a helical spiral, illustrated here in three dimensions; (g) living *Rhabdopleura*, a benthic pterobranch with zooids, shown extended and retracted, within their tubes – the collagen skeleton, like that of graptolites is made of fusellar rings; (h) Reconstruction of part of a *Climacograptus*, showing zooids, modelled on those of *Rhabdopleura*, 'painting' on cortical tissue.

process takes place in the living pterobranch *Cephalodiscus*. Here, as we have seen, the skeleton forms a hemispherical mound, with very many long protective spines projecting from it. It is pitted with many cavities in which the zooids reside. These, unlike the attached zooids in graptolites, are free-living. They look like very small tadpoles, and when 'working' they sneak out of their holes, paint on the collagen of

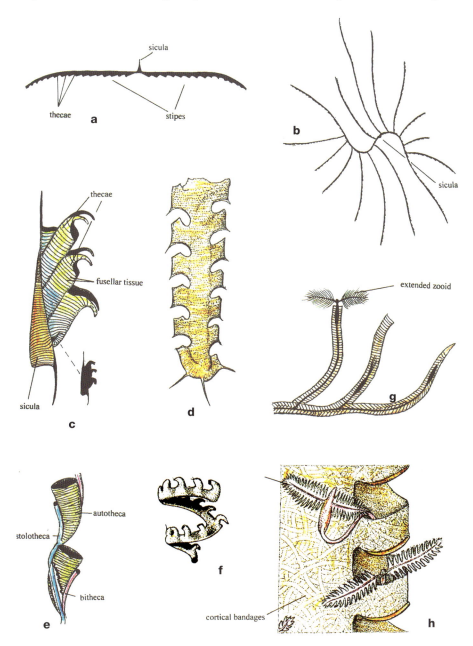

which the skeleton, including the spines, is constructed, and finally sneak back in again. This kind of skeletal construction is typical of all graptolites, and we can keep it in mind as a model when briefly considering the geological history of the group as a whole.

There are several groups or Orders of graptolites, of which only two are important. These are the Dendroidea (Upper Cambrian–Carboniferous), which we have so far only encountered in passing, and the Graptoloidea (Lower Ordovician–Lower Devonian) of which *Didymograptus* and *Saetograptus* are representatives. There are some other rare, sessile forms also which we need not consider further.

The oldest order comprises the dendroids, and it is from these that the graptoloids originated. Most dendroids are benthic; they were rooted to the sea floor and formed shrub-like masses with many stipes (Fig. 9.7 (1)). As Wiman showed in the late nineteenth century, and as Kozłowski so conclusively confirmed, dendroids have three kinds of thecae, rather than two (Fig. 9.6e). The stolothecae (homologous to the graptoloid prothecal range) form continuous branching closed chains, and it is from these that the so-called autothecae (equivalent to the graptoloid metathecae) arise. These are quite large and robust and probably housed feeding zooids. The third kind are known as bithecae; they are thin tubes arranged in a particular pattern for each genus, and probably contained reproductive zooids, fed by the other members of the colony through an internal system of connections. These three kinds of thecae were formed of fusellar tissue, but the cortical tissue was much thicker, and ensured that the colony was firmly rooted and held upright.

Most dendroids, such as *Dendrograptus* and *Dictyonema*, were moored to the sea floor. They fed on particles raining down from the upper waters of the sea. Although they survived until the Carboniferous, they were never very numerous except locally, and this rather passive mode of feeding was surely somewhat restrictive. Might there not be merit in colonising the realm of the plankton, where so much more food was

Opposite **9.7** Change in shape of graptolite colonies through time. The sketches of graptolites are placed, as far as possible, at their correct stratigraphic level. (1) The dendroid genus *Dictyonema*, rooted to the sea floor, gave rise to the first planktonic dendroid; (2) *Rhabdinopora* (L. Ord.), living suspended from seaweed. Descendent forms were free-floating and had several stipes, e. g. (3) *Staurograptus* (L. Ord.); (4) *Tetragraptus* (L. Ord,) and the scandent (5) *Phyllograptus* (L. Ord.). Stipes reduced to two in number as in (6) *Isograptus* (L. Ord.) and (7) *Didymograptus* (L–M. Ord.). This latter is a long-ranged genus, but individual species such as *D. murchisoni,* illustrated here, have short ranges and are useful in biostratigraphy. (8) *Nemagraptus* defines the base of the U. Ord. *Dicellograptus* (M–U. Ord.) is also stratigraphically useful; (9) *D. elegans* (U. Ord.) and (12) *D. anceps* (U. Ord.) are zone fossils, the latter defining the highest Ordovician zone; (10) *Dicranograptus clingani* (U. Ord.) is likewise

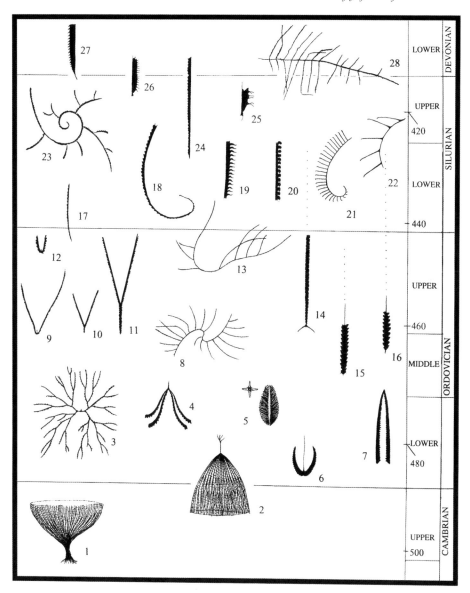

a zone fossil and occurs along with (11) *D. ramosus* (U. Ord.); (13) *Pleurograptus linearis* (U. Ord.) is also an important zone fossil; Scandent genera (14) *Climacograptus* (U. Ord.–L. Sil.); *C. bicornis* (U. Ord) illustrated; (15) *Orthograptus* (M. Ord.–L. Sil.) and (16) *Glyptograptus* (L.Ord.–L. Sil.) become dominant and continue into the Lower Silurian after the first monograptids appear – these have thecae on one side of the stipe only; (17) *Atavograptus cryx* (lowest Sil.); (18) *Coronograptus cyphus* (L. Sil.); (19) *Stimulograptus sedgwicki* (L. Sil.); (20) *Campograptus lobiferus* (L. Sil.); (21)*Rastrites peregrinus* (L. Sil.); (22) *Rastrites maximus* (L. Sil.); (23) *Cyrtograptus* (U. Sil.); (24) *Pristiograptus dubius* (U. Sil.); (25) *Saetograptus chimaera* (U. Sil.); (26) *Saetograptus leintwardinensis* (U. Sil.); (27) '*Monograptus*' *uniformis* (L. Dev.); (28) *Abiesgraptus multiramosus* (L. Dev.).

available? This is precisely what one dendroid genus, *Rhabdinopora*, actually did (Fig. 9.7(2)). Whereas this is very similar to the rooted, benthic genus *Dictyonema*, the cortical tissue is much thinner, and the root was, in all probability, attached to floating seaweed. From this arose the earliest graptoloids, which broke free of attachment, and floated freely in the water.

We shall first briefly consider the evolutionary history of the graptoloids, which forms the basis for their biostratigraphical application, but also reflects various kinds of adaptation for feeding. The early Ordovician graptoloids were quite diverse. Such genera as *Radiograptus* and *Dichograptus* retained many stipes from the ancestral dendroids, but these became more equally spaced and regularly arranged, lying in a horizontal plane. Sometimes these were connected centrally by thin web-like plates, which would give extra stability. The bithecae disappeared, their function being taken over by the zooids that inhabited the metathecae; there is one genus in which the bithecae open into the metathecae, rather than externally.

Although these large forms (anisograptids) were common in the early Ordovician, there was an overall reduction in stipe number. *Tetragraptus* (Fig. 9.7(4)) has four stipes, in different species extending in a plane, drooping downwards and turning up again, but soon two stipes became standard, as in our example *Didymograptus extensus*. Different kinds of two-stiped graptoloids persisted until the end of the Ordovician, providing a basis for biostratigraphy. The earlier forms tended to have the stipes extended outwards, downwardly drooping or parallel-sided like a tuning fork, while those of the later Ordovician were more commonly inclined upwards, as in the stratigraphically important genus *Dicellograptus* (Fig. 9.7 (9, 10, 12)) and its relatives. Sometimes these grew extra stipes laterally. In the middle Ordovician there arose for the first time 'biserial scandent' genera, in which the two stipes grew back-to-back, the thecae facing in opposite directions (Fig. 9.6d, 9.7 (14–16)). Some of these scandent types were of considerable morphological complexity, especially in the first-formed thecae round the sicula. The retiolitids are a long-lived group in which the first few thecae are of normal form, but with a meshwork of secondary peridermal tissue growing outside of these. Whereas in the Lower Ordovician the thecae formed simple tubes, those of the later Ordovician were of much more complex and twisted form, the various types of which are very useful in recognising different genera. Whereas biserial scandent graptoloids persisted well into the Lower Silurian, the dominant Silurian and Lower Devonian genera were the 'uniserial scandent' monograptids, such as our example *Saetograptus* with thecae on one side of the stipe only (Fig. 9.6c, 9.7(25)). Up till now graptoloid stipes had been mainly straight, but many of the Silurian forms are coiled, forming logarithmic or Archimedean spirals. The well-known middle Silurian genus *Cyrtograptus* (Fig. 9.7 (23)) is a spiral form with lateral straight or slightly curved cladia arising from various parts of the spiral, whereas *Monograptus turriculatus* forms an elegant, conical helical spiral. The forms of the thecae vary dramatically in successive Silurian species; some are straight, others hooked, lobate, laterally twisted or taking other forms. Monograptids persisted into the lower Devonian; they went on evolving to the last and there is no real reason why they should have become extinct.

Graptolite feeding and ecology

The large early graptoloids lived in the upper waters of the sea, feeding on smaller plankton. Experimental modelling has shown that if the stipes are angled in a particular way, these graptoloids will rotate when sinking, thus sweeping out an area of the water column and acting as a kind of harvesting machine. But it could not go on sinking for ever. In order to rise up again (harvesting while it did so) some buoyancy system would be required. One possible mechanism would involve symbiotic algae within the bodies of the zooids. During the day these would photosynthesise, releasing oxygen bubbles, thereby giving lift. At night photosynthesis ceased, the graptoloid would sink, probably no more than a few metres, and would rise, spiralling upwards, when photosynthetic activity began again next morning. This colony would feed both when rising and sinking. The S-shaped *Nemagraptus* (Figs. 9.6b, 9.7(8)) with its lateral branches or cladia presumably functioned in a like manner.

Other versions of this kind of harvesting arose later, after the extinction of these large anisograptids. In the Middle and Upper Ordovician, for instance, there arose V-shaped and Y-shaped graptoloids (*Dicellograptus* and *Dicranograptus*) (Fig. 9.7 (9–12)), which may have slowly spun round and round in the current. Likewise in the Silurian, the helically coiled *Monograptus turriculatus* (Fig. 9.6f) probably functioned in this way, as did the large Middle Silurian *Cyrtograptus* (Fig. 9.7(23)) which formed a large low-angle helical spiral, but also had extra branches ('cladia'), which would increase the harvesting area, and would further induce rotation of the colony in weak currents. Sometimes the nemas of several rhabdosomes are found tangled together so that a kind of supercolony (synrhabdosome) is formed. These are very rare in any case, but at least partially because they were the subject of some rather fanciful reconstructions in the earlier years of last century, most palaeontologists tended to dismiss them. Yet in recent years, some bedding planes in the Spanish Ordovician have been shown to be covered with them, and in these the nemas are truly linked. It is quite possible that these linked colonies, with their overall increased surfaces, also acted as harvesting machines. Some graptoloids, especially the scandent forms, may simply have floated passively in the water, but some mobility may have been given to the colony by feeding currents produced by the zooids alone. For over 80 m.y. the graptolites dominated the zooplankton, and developed many adaptations for feeding and floating; those noted above are only a few examples, and there is very much that is still not known. The life and ecology of these fascinating marine animals will continue to intrigue us for many generations to come.

Graptolite biostratigraphy

Graptoloids have been used for stratigraphical correlation since the time of Lapworth. There are several reasons why they are excellent zone fossils, notably that they were planktonic and widely dispersed, they were not confined to narrow, temperature-dependent latitudinal belts, and they evolved very rapidly with a high rate of turnover of different genera and species. Where present in black shales, this rapid rate of change provides an excellent basis for biostratigraphy, and many graptolite biozones define time periods of no more than half a million years. This is not the case for all graptoloids, however, and some genera are too long-ranged to be of stratigraphic value. There are two further problems in using graptolites for biostratigraphy, the first being preservability, the second provinciality.

Preservability. Graptolites of all kinds are fragile. They are only normally preserved in dark, fine-grained shales, which accumulated in stagnant bottom waters where scavengers were deterred by low oxygen levels. Commonly, in such environments, they are squashed flat, and there may be problems in precise identification. As we have seen, occasionally they are present and preserved in three dimensions in fine limestones. But otherwise they are not often found in nearshore or coarser sediments, and mixed graptolite and shelly associations are relatively uncommon. Where they do occur, of course, it is possible to correlate precisely between open-sea and nearshore sediments. But more commonly, throughout the Lower Palaeozoic, there are critical sections where graptolite shales interdigitate with sediments containing trilobites, brachiopods and other 'shelly faunas', the result of fluctuating sea-levels. In such cases the alternate bracketing of graptolites with shelly fossils allows a high degree of mutual correlation; the zones based on graptolites and those based on trilobites and graptolites can be accurately matched.

Provinciality. Some graptolite species are widespread, others occupied a more limited geographical area and are useful for local correlation only. In the early Ordovician, graptolite faunas were much the same all over the world. Somewhat later two faunal provinces became established, and in these the faunas are not closely similar. If we plot the occurrences of these two kinds of graptolite faunas on palaeo-geographic maps for the Ordovician and Silurian, we find they fall into two broadly latitudinal belts, a tropical 'Pacific province' and a cooler-water 'Atlantic province'. Our southern Scottish graptoloids belong to the former, and are more similar to those of North and South America than they are to those of England and Wales and most of the continent. But again, interdigitating Atlantic and Pacific faunas, together with the rare mixed faunas, supply a solution to the problem. It is becoming clear that as with so many of today's plankton and nekton, individual water masses have their own specific graptolite faunas, and moreover that different species and genera inhabited different depths; some graptoloids appear to have been confined to the oxygen-minimum zone. Whereas the 'Atlantic' and 'Pacific' provinces were well marked throughout most of the Ordovician, the advent of the great glaciation at the end of the Ordovician put an end to the cooler Atlantic province. Recovering in the Silurian, the graptoloids spread all over the world, though later, and for a time, the two provinces were re-established. By the end, in the later Silurian and until their extinction in the early Devonian, graptolites were confined to tropical seas alone.

So, although there are problems for correlation, they have not proved insuperable, and the relative timescale based on graptolites is robust. As we have seen, some graptolite zones define time periods of half a million years or less. Others are somewhat less finely resolved. But even at the coarsest possible level of biostratigraphical resolution, the overall sequence of graptolite faunas proves valuable. It would not, for example, be possible to confuse a Lower Ordovician anisograptid assemblage with a fauna of biserial and uniserial scandent graptoloids from the early Silurian. There is no doubt that graptolites, as a whole, will retain their stratigraphic value and their biological and ecological interest for the foreseeable future.

CHAPTER 10

The Moffat Shales — graptolites, eruptions and glaciations

Extending north-eastwards from Moffat towards Peebles and beyond are the highest hills in central southern Scotland, attaining 840 m at Broad Law. Some, to the east of the M74, are cut by deeply incised ravines formed where the softer black Moffat Shales have been preferentially eroded. In some of them, the graptolitic shales are so tectonically smashed that little ordering of the sequence can be determined, but in others, the rocks are relatively undamaged and the stratigraphic sequence remains intact. Here we consider a few of these localities, for they all contribute to the story of the sedimentary environment, climatic changes, graptolite evolution and the subsequent deformation related to Iapetus closure. The most instructive of all remains that of Dob's Linn (Fig. 10.1) only a few hundred metres from Birkhill Cottage, where Lapworth stayed between 1872 and 1878 and where he produced his largest and most perfect geological map (Fig. 10.2).

Dob's Linn

On the north side of the A708 road, some distance southwest of Lapworth's cottage, a grassy slope leads down to the burn which becomes the Moffat Water. To the northeast is the main ravine, with black shales exposed on both sides. Those on the southern face are highly deformed and useless for stratigraphy, but those of the northern face present a fine succession of gently tilted, almost undeformed shales.

Ascent brings one to a point affording a spectacular view of a waterfall from which Dob's Linn takes its name. At the bottom of the upper cascade is a shallow cave, where a fugitive Covenanter, Halbert Dobson, on the run from English soldiers, hid for several weeks in the 1690s. The soldiers who searched for him never found him. Dobson's Linn became abbreviated to Dob's Linn, although the older form, Dobb's Linn, is sometimes used.

10.1 The waterfall at Dob's Linn, which falls over a vertical face of hard Silurian greywacke. To the left are softer black shales of Lower Silurian age, replete with graptolites.

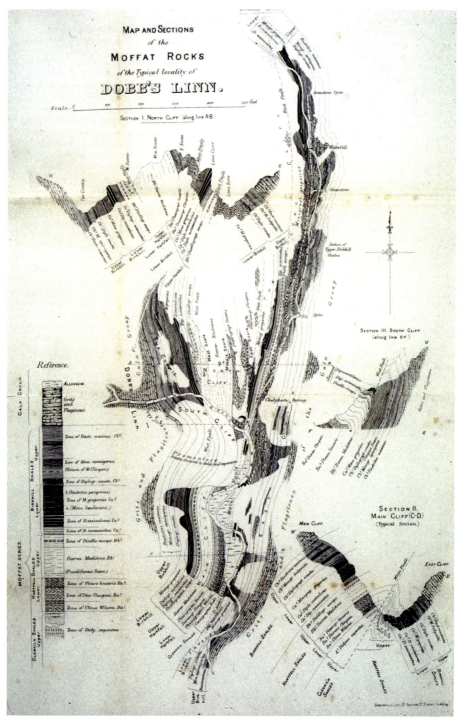

10.2 Lapworth's original map of Dob's Linn from 1880.

On the western side of the ravine are almost vertical, slightly flexed layers of Lower Silurian black shales. Closer to the waterfall, the first greywackes appear above the shales. In the near-vertical face of one of these greywacke units (to the right of the waterfall) is a curving groove where a pebble, impelled by the turbidity current which brought in the greywacke, excavated a scratch in the underlying mud immediately before deposition of the sand. It is an excellent example of a tool mark, as described in Chapter 5.

Biostratigraphical division

The shales and greywackes at Dob's Linn range from the base of the Upper Ordovician up into the Lower Silurian. Figure 9.7 shows the overall succession and, specifically, how the shapes of graptolite rhabdosomes changed through time. Rather than using the Series and Stages into which the Ordovician and Silurian Periods have been subdivided, we shall simplify by referring only to the Lower, Middle, and Upper Ordovician and Silurian.

No fewer than fifteen distinct graptolite zones can be distinguished within the shales that compose the bulk of the sequence. These zones (normally the smallest divisions used by biostratigraphers) are based on the appearance and disappearance of graptolite species. Most of the zones are believed to represent less than a million years' worth of sedimentation. The zones are in practice referred to by their specific names, as a kind of shorthand. Thus the Upper Ordovician *Dicranograptus clingani* Zone is known simply as the *clingani* Zone and the *Rastrites maximus* Zone as the *maximus* Zone. Although a zone is defined on the vertical range of a specific species, it commonly contains several other species, some of which may be longer ranged than the species on which the zone is defined. However, the zones defined by Lapworth, although still of local use in Scotland, have been superseded by a revised system of zonation that is globally valid. For example, *Rastrites maximus* does not occur outside southern Scotland, and is therefore no use for correlating Silurian strata anywhere else. In the 'global standard' for the Ordovician and Silurian Periods the Scottish *maximus* Zone is now accommodated within that of the more widely recognised *guerichi* Zone.

Rock sequence at Dob's Linn and its interpretation

The oldest rocks at Dob's Linn are the black, cherty Glenkiln Shales, seen in the first exposure south-west of the Main Cliff, and shown in Lapworth's original (Fig. 10.2). However, they are poorly exposed here and are much better at the type locality of Glenkiln Burn, in Ayrshire. Next up are the Lower Hartfell Shales, which are black and rich in graptolites, with three zones represented. In these the diversity, relative abundances, and mode of preservation of the graptolites fluctuates dramatically through time. Thus the *clingani* Zone bears, as its zone fossil *Dicranograptus clingani*, a small Y-shaped graptolite with a short biserial shank, but there are many other species. The large *Dicranograptus ramosus* (Fig. 9.5a), the zone fossil of the overlying zone of *Pleurograptus linearis*, is a large, thin, spindly form with lateral branches (cladia), easily broken (Fig. 9.7(13)). It often occurs with swarms of *Orthograptus truncatus* (Fig. 9.4). While further details of the graptolite succession need not concern us, what is important is the sharp change from these black shales to the grey-green and almost unfossiliferous (Upper Hartfell) shales or Barren Beds, which pass up almost to the very end of the Ordovician. Only two thin horizons of 'normal' graptolite shale are found within the main part of this sequence, belonging to the *complanatus* Zone. But towards the top of the Ordovician are some remarkable thin intercalations of black shales of the *anceps* Zone, in which various graptolite species occur, alternating with thin barren strata. Then black shales take over again, passing up into the Silurian (Birkhill) shales, which persist through several zones until the greywackes come in. These mark the arrival of the giant submarine sedimentary fans to the Dob's Linn area, thereby completely altering the nature of sedimentation.

Meta-bentonites

The black coloration of the typical Moffat shales is due to the relatively high content of carbon (graphite) particles accompanying the minute clay crystals. The most plausible explanation is that the carbon came from the decay of floating algae (seaweed) that otherwise left no coherent fossil record.

The Dob's Linn shales, however, are interleaved with one hundred and thirty-five pale grey to whitish unfossiliferous layers that vary from about 1 to 50 cm in thickness (Fig. 10.3). From their mineralogy and chemistry it is clear that these represent volcanic tuff. The inference is that the tuffs were blown high into the atmosphere by explosive eruptions and carried by the winds before raining down onto the Iapetus Ocean. Thus the succession comprises two highly contrasting types of strata, namely the black shales, representing extremely slow accumulation of tiny clay particles far from land and incorporating graptolite fossils, and 'ash' layers generated by volcanic eruptions. Whereas the shale layers are reckoned to have taken tens of thousands of years to accumulate, each of the tuff layers probably formed over a mere few years.

The tuffs probably commenced as friable particles of pumice. However, they have been so altered by subsequent reaction with seawater and tectonic compression that some imagination is required to visualise how they may have appeared when they first came down from the sky. The word 'pumice' refers to a volcanic glass full of cavities, similar to the material used as bathroom toiletries, and can essentially be

10.3 Finely-bedded Moffat shales at Dob's Linn with a pale-coloured layer of altered tuff ('meta-bentonite') a few centimetres to the right of the ball-point pen.

thought of as a solidified froth. When initially blasted from their volcanic conduits the pumices would have been incandescent molten blobs of magma entrained in a high-velocity gas discharge and carried up to heights that could have reached tens of kilometres. Gases formerly dissolved within the magma escaped to produce the myriad bubbles as pressures reduced when the magmas neared the surface. (The gas separating from the melt can be likened to the frothing of a bottle of Guinness as the pressure is released on opening.) As they lost both heat and gases, the pumice blobs would have congealed to a natural glass. So we may imagine the pumice as having occurred as sub-rounded particles, ranging from pea- to orange-sized, and consisting of frothy glass capable of floating on water. Accumulation of millions of these, descending from a big eruption column, would have formed great rafts of floating pumice. Whereas some may have sunk without delay, much is likely to have floated for variable periods before becoming waterlogged and sinking to the ocean floor.

To envisage the probable scenario we can turn to sailors' accounts of volcanic fallout at sea, like those resulting from the apocalyptic eruptions of Krakatoa between Java and Sumatra between May and August 1883. Thus the steam-ship *Annerley* reported steaming through a sea of pumice stone while the *Loudon* found itself sailing through a sea of pumice debris about a foot thick, with the log noting that the floating pumice fields could scarcely be seen across. After settling on the ocean floor, compaction of the pumices and hydration reactions with seawater would have commenced. The glasses, ephemeral materials at the best of times, would largely have been replaced by minerals of the clay family. As a consequence of the deformations that marked the end of Iapetus, the tuff layers underwent intense compaction, and their original thicknesses will have been greatly reduced. Ancient 'ash' layers are commonly described as 'bentonites', but those at Dob's Linn are so altered from their original condition that they are referred to as 'meta-bentonites', their volcanic origins only being evidenced by their distinctive mineralogy and chemical composition.

While several large eruptions may be expected each century, really catastrophic eruptions of what have been called super-volcanoes, involving the spewing of hundreds and even thousands of cubic kilometres of disseminated magma into the atmosphere, are mercifully rare events that tend to be separated by tens of thousands of years. The whole succession at Dob's Linn is estimated to have taken about 25 m.y. to accumulate so, had the volcanic episodes occurred at regular intervals, there could have been gaps of 150,000 to 200,000 years between eruptions.

The most horrendous eruptions in historic times are all associated with supra-subduction volcanoes. Each century there are several large eruptions capable of ejecting magma measurable in terms of cubic kilometres, much of this being as 'ash' particles. The finest of these can be blown high into the upper atmosphere and become disseminated on a global scale. Questions arise as to what sort of volcanoes were involved in the formation of these Ordovician–early Silurian meta-bentonites and in what geographic settings? The answers have to be extremely speculative, bearing in mind that the volcanic sources could have lain hundreds or even thousands of kilometres distant from the outcrops under discussion. Nonetheless, we may reckon that most were supra-subduction arc volcanoes (as described in Chapter 4) accompanying the closure of the Iapetus.

Most of the meta-bentonites probably arose from the relatively silica-rich (silicon dioxide) magmas referred to as dacites and rhyolites, commonly associated with big eruptions. Some, however, have chemical characteristics that are rarely, if ever, seen in the supra-subduction volcanic arc associations. A number of the meta-bentonites have trace-element signatures denoting an origin from unusually alkaline magmas, the so-called 'peralkaline rhyolites'. We have already encountered these peralkaline rhyolites in Chapter 5 with regard to the Tweedsdale 'Wrae volcanics'. Such magmas are untypical of supra-subduction arc volcanoes, and are more commonly developed within large oceanic intra-plate volcanoes. Our best guess is that the peralkaline meta-bentonites came either from one or more oceanic island volcanoes in Iapetus or possibly from volcanoes in one of its back-arc basins.

It appears that the successive rains of pumice affected the graptolite faunas. Careful collection, layer by layer, through the black shales interleaved with the meta-bentonites has shown that the latter are indeed coupled with extinctions and changes in the graptolite fauna, although much more work needs to be one on this subject. If the ocean waters were indeed blanketed by pumice for several weeks or months after eruptions, such extinctions would be hardly surprising.

Hirnantian glaciation

The remarkable changes in the rock record at Dob's Linn could not have been interpreted by Lapworth. We know now that they were related to climate change. The black Lower Hartfell Shales were fine-grained, anoxic, ocean-floor muds, the finest fractions of material derived from the land, deposited in deep water. They are not, however, entirely homogeneous, but show distinct variations and minor cyclicity. It is now well established that a period of global warming occurred in the late Ordovician (the Boda Event), followed by a temperature fall. The abrupt change to the pale Barren Mudstones denotes the onset of colder conditions in the later part of the Ordovician, with oxygenated bottom waters sinking down from polar regions and ventilating the sea floor. Anoxic conditions returned briefly during the *complanatus* Zone, but otherwise, during this long period of cooling, productivity in the upper waters of the sea may have been reduced, judging by the lack of carbon in the sediments. In any case conditions were not right for the preservation of the graptolites.

There are clear indications that at the end of the Ordovician (*c*.445 Ma) the Earth experienced a severe glaciation, marking the close of a long warm interval that had lasted since the previous ice age in the late Precambrian. Global sea-levels would have oscillated in harmony with the growth and shrinkage of the ice cover (giving a contemporary aspect to these Palaeozoic climatic changes). The southern continent of Gondwanaland settled over the South Pole, developing an ice-sheet, which appears to have been commensurate with that of the modern sheet over Antarctica.

The effect on faunas of this sharp glacial episode (the 'Hirnantian ice age') at 439 Ma towards the close of the Ordovician was severe. The uppermost Hartfell Shales were being deposited in the late Ordovician and the alternation of black and grey shales testifies to dramatic swings between colder and warmer conditions. The Hirnantian glaciation ended with a rapid deglaciation and a return to anoxic conditions as represented in the Birkhill Shales of the latest Ordovician and lowest Silurian. This pattern of sedimentation persisted until the incoming of the Gala greywackes. The Hirnantian glaciation, drastic though it was, had lasted less than a million years. It was possibly the shortest really extensive glacial period in geological history.

Boundary stratotype

The boundaries between the Periods have to be precisely defined at some specific point somewhere in the world, by a 'boundary stratotype'. Some years ago, the International Committee dealing with the Ordovician/Silurian boundary decided that Dob's Linn was the best place for the boundary stratotype, at least for continuous successions in black shales (additional reference sections are required for shallow-water successions). This did not please all geologists, notably the Canadians, who felt that their successions in Anticosti Island were better in all respects, but the 1985 decision of the Committee was binding and is now accepted. The boundary stratotype lies within a continuous succession of highly graptolitiferous black shales, at the base of the *acuminatus* Zone, 1.6 m above the base of the Birkhill Shales. One may wonder how many geologists have been photographed with their finger pointing at this critical boundary!

Dob's Linn is, however, only one of many ravines in southern Scotland, and they all tell different parts of the story. We shall consider just three particularly interesting examples.

Hartfell Score

North of Moffat, along the A701 as far as the remarkable deep bowl of the Devil's Beef Tub, the mountains seen to the east are dominated by Hart Fell (808 m). Some distance below its summit, Hartfell Score is a deeply eroded gash, providing one of the finest of all the exposures of the Moffat Shales (Fig. 10. 4). The Score itself used to be visited for the medicinal waters of the chalybeate spring, which though still present, is no longer used as a spa. The southern face of the Score is severely tectonically disturbed and most of the central section is eroded into spectacular pillars, pinnacles, sheer cliffs and gullies, too complex to interpret geologically. Lapworth (1880), astute observer that he was, noted: 'Nowhere to the north of the Moffat valley are the dark shales so greatly folded and shattered as in this locality; but on the other hand, nowhere are the fossils so prevalent throughout the beds...'. This remains true today, for the graptolite specimens lying around on detached slabs of shale are of excellent quality. However, the range of strata exposed at Hartfell Score is much less than at Dob's Linn, and Silurian beds (Birkhill Shales) occur only in the highly deformed sections in the southern cliffs where the graptolites are hardly identifiable. The best section is towards the western end of the Score where the shales are unfolded but are repeated by thrusting. Here are poorly fossiliferous Glenkiln Shales, above which lies a splendid sequence in the black Lower Hartfell Shales, yielding well-preserved, and eminently collectable specimens.

10.4 Hartfell Score, looking eastwards.

Craigmichen Scaurs

This remarkable section lies to the south of the A702 road, reachable from Selcoth Farm. The Scaurs are cut into the south-eastern flank of Capel Fell. What is most striking is the scale of the exposure, as it is about a kilometre long and five times the height of that at Dob's Linn. Although the lower part presents a sheer cliff, the higher parts are more accessible and yield a profusion of graptolites. The north-western face exposes Ordovician strata with radiolarian cherts, Glenkiln Shales and the Lower and Upper Hartfell Shales, repeated by significant thrust faults. Whereas in most respects the geology is typical of the Southern Uplands, the beds representing the Ordovician zones are twice as thick as at Dob's Linn, indicative of much higher rates of sediment deposition. Clearly the succession exposed here was deposited in a quite different part of the Moffat Shales basin that was subsequently juxtaposed by faulting to a position closer to Dob's Linn. But where Dob's Linn and Craigmichan Scaurs originally lay with respect to each other we have no way of telling.

Thirlestane Score

Some distance south of Tibbie Shiel's Inn, on the western shore of St Mary's Loch, is the high valley of Thirlestane Score, where a fine sequence of Silurian shales is exposed. It looks very different from most of the other exposures in the Moffat Group, for although black shales are present at the base, those forming the upper part of the sequence are pale pinkish-purple (Fig. 10.5). Lapworth regarded the black shales as belonging to his *Monograptus spinigerus* Zone, while the overlying purple shales form the type locality of *Rastrites maximus*. The latter is a strange and relatively large graptolite, in which the thecae are very long and isolated one from another (Fig. 9.7(22)). Originally it may have had a spiral form, but it is never found complete, and it seems that these delicate graptolites broke up after death and only sections of the whole stipe remain. It is a local graptolite, not found outside southern Scotland, and the *maximus* Zone, though it can be used locally, is now regarded as part of the globally recognised *Monograptus guerichi* Zone. The underlying beds of Lapworth's *Monograptus spinigerus* Zone now belong to the *Monograptus halli* and *Stimulograptus sedgwicki* zones.

The change from the highly fossiliferous black shales of the lower two zones to the pale purple shales of the overlying *guerichi* Zone is dramatic, and there must have been a good reason for it. The likely answer is that ocean current systems changed. Small animals among the zooplankton eat the phytoplankton, the primary producers. The nutrients essential for life tend to disappear from the upper waters of the sea in the dead bodies and faecal pellets of the plankton and of the larger predators

10.5 Thirlestane Score, looking westwards. The pinkish purple colour of the upper part of the Silurian shale succession (*guerichi* zone) is very distinct.

that prey on them. Nitrates and phosphates are thereby lost to the upper waters and accumulate on the sea floor. In today's oceans, high planktonic productivity is largely sustained by upwelling, where currents rising from the depths bring up these lost elements again so that they are once more available for the plankton. That is why fish, the end members of the oceanic food chains, are (or rather were) plentiful along continental margins such as the Peruvian coast, and the Newfoundland Banks and Iceland, where upwelling occurs on a major scale. It seems clear enough that the black graptolitic shales of southern Scotland and elsewhere were deposited from waters of high productivity. Such is the case, at Thirlestane Score, for the *halli* and *sedgwickii* zones. Then an abrupt change occurred, probably through a shift in oceanic current systems that could have had a variety of causes. Whatever happened, the area of what is now Thirlestane Score was no longer in a productive part of the ocean and elements sinking to the sea floor were not recovered. But there is a curious twist to this story. For even within the purple shales there are some strata with graptolites, including *Rastrites maximus*. These always occur in association with meta-bentonites, and careful sedimentary logging on a very detailed scale allowed the following interpretation of what happened. Each of the newly erupted tuffs supplied iron, aluminium, or trace metals to the upper waters of the ocean. These elements were taken up by phytoplankton, which underwent a bloom and, in turn, by the graptolites. Just above the meta-bentonites, juvenile graptolites are abundant, for although there must have been enough to eat, the ambient environment in the early stages of a plankton bloom is often stressed, and juvenile mortality was high. Later, a more mature ecosystem was attained, with large and full-grown graptolites being present. But soon after, productivity decreased, and in a relatively short while the absence of the essential metals put an end to the short-lived phytoplankton and graptolite blooms. This scenario was repeated on no fewer than eight occasions at Thirlestane Score. Some scientists have suggested that productivity in the less 'good' regions of the ocean today could be stimulated by seeding them with iron and other elements. The Thirlestane Score experience suggests that it could work, but to tamper with Nature's stable systems is not necessarily a wise course of action.

CHAPTER 11

Tsunami!

The Southern Uplands Fault traverses Scotland and across the Irish Sea into Northern Ireland. Over most of its length Upper Palaeozoic sediments are thrown against the Ordovician of the Northern Belt. Except at Girvan, there is no trace of the continental shelf which must have lain to the north; it has either been 'lost' by faulting or covered by younger rocks. Yet there is tantalising evidence of the former continental shelf, from an indirect source. For at several sites in the Northern Belt there are shallow-water fossils (trilobites, brachiopods, gastropods, bivalves, and others) in great abundance that can only have come from the original shelf, and have slipped down into deep water. Such fossils were first found by a Survey officer, R.L. Jack, in 1868, and collections were made in 1869 by the Survey's experienced collector, A. McConochie, from various localities, in the Leadhills district, west of the Clyde Valley, and Kilbucho and Wallace's Cast, east of Crawford and Abington, in the Borders. Over thirty species were listed by Peach and Horne in 1899, and the fauna was recognised as Upper Ordovician, a determination sustained by more recent work.

What happened at Kilbucho?

Of the various localities, the section at Kilbucho is perhaps the most instructive and easiest of access (Fig. 11.1); many of the others are remote from any road, and difficult to find. The fossils therein, though occurring at all these other places, are here referred to as the 'Kilbucho fauna'. The section was rediscovered in 1973, and then in the mid-1980s a team from Scottish universities and Durham revisited Kilbucho and various other sites with the intention of establishing the diversity of the fauna prior to determining the age as precisely as possible, and to compare it with the faunas of other sites of equivalent age. Moreover, these geologists wished to establish how these shallow-water marine invertebrates had come to be entombed in sediments deposited in deep water.

11.1 The old quarry near Kilbucho, between White Hill (N) (right) and Culter Fell (S) (left). Cecilia Taylor for scale.

The Kilbucho locality lies in a long valley south of Biggar, on the southern slopes of White Hill (434 m) and north of the great mass of Culter Fell (748 m) (Fig. 11.1). The fossiliferous sequence lies about a kilometre west of the ancient ruined church of St Bega, from which Kilbucho takes its name, on the lower slopes of White Hill (NT 056336). Here, the lowest part of the sequence is exposed in a farm quarry, and consists of thinly bedded unfossiliferous greywackes, belonging to the Kirkcolm formation. They are nearly vertical, and sole markings on their lower surfaces indicate a northward younging. Above lies a conglomerate, which is of the greatest interest (Fig. 11.2).

The contact between the greywacke and the overlying conglomerate is unexposed, although at other localities it can be seen clearly, and is erosive with the sharp base of the conglomerate cutting down into the underlying greywacke. The Kilbucho conglomerate is unsorted, and the clasts within it are supported by the matrix. It consists of angular pebbles of white quartzite, vein quartz, chert, assorted igneous and metamorphic rocks and mica flakes. There are also individual fossils including brachiopods, corals and crinoids which occur together with rounded pebbles (up to 30 mm) of highly fossiliferous, decalcified mudstone, with undeformed shells. These must have lithified before their erosion from some pre-existing source.

The conglomerate grades up into a brown mudstone within which fossils are very abundant, though their original shell material has been leached away so that they

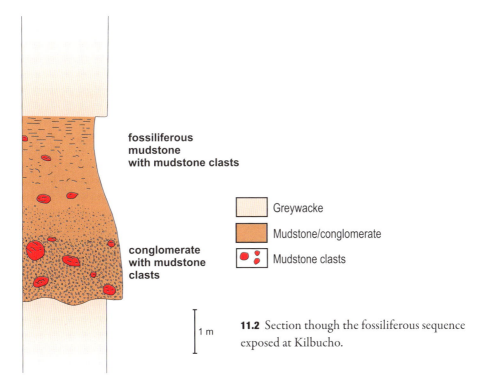

fossiliferous
mudstone
with mudstone clasts

Greywacke

Mudstone/conglomerate

Mudstone clasts

conglomerate
with mudstone
clasts

1 m

11.2 Section though the fossiliferous sequence exposed at Kilbucho.

are preserved as moulds of the external and internal surfaces (Figs. 11.3, 11.4). They often lie across the bedding, and may be crushed or distorted. Not uncommonly the fossils are broken, having evidently been cracked during transit, but the pieces held together like crazy-paving. Fossiliferous fragments of mudstone occur also within the matrix of the overlying mudstone. These are identical to those in the conglomerates, though smaller.

The fossils thus occur in three ways; as separate individuals within the conglomerate; as undistorted moulds within the fossiliferous mudstone clasts both in the conglomerate and in the overlying mudstone; and as moulds, often distorted, crazed and broken, within the mudstone. They are relatively sparse within the conglomerate, best preserved in the mudstone clasts, and most abundant within the mudstone.

The top of the sequence is unexposed, though a little way above the quarry with the mudstone, greywackes are present. The fossiliferous sequence, sandwiched within the greywackes, is probably no more than three or four metres thick. But before we attempt to interpret it, we should consider other localities in the Northern Belt where fossils occur, to see if the same kind of sequence is represented and to consider the nature of the fauna.

The valley of the Wandel Water lies eastwards of the hills above Crawford and Abington, in the Clyde Valley. It is of considerable geological interest, displaying

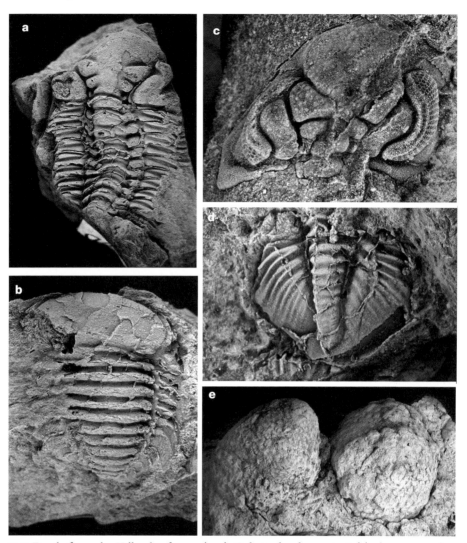

11.3 Fossils from the Kilbucho fauna: (a–c) *Calyptaulax brongniarti* (a) almost complete trilobite ×2, (b) head ×4, (c) tail ×4; (d) *Remopleurides* sp. ×4; (e) *Mastopora* sp., an alga which indicates a very shallow water origin for the fauna ×3.

ocean-floor volcanic rocks (hyaloclastites), overlying red radiolarian cherts and mudstones (from which conodonts have been collected), graptolitic shales, and greywackes. Within the greywackes, close to the curious erosion feature known as Wallace's Cast, lies a patchily exposed sequence of fossiliferous rocks, virtually identical to those of Kilbucho. However, the conglomerate here is more degraded than that of Kilbucho, and fossiliferous mudstone clasts are found both within it and in the overlying mudstones. It is, in other words, a more or less identical sequence, carrying the same fauna.

11.4 Fossils from the Kilbucho fauna: (a) *Kilbuchophyllia clarksoni,* solitary coral ×2; (b) *Kilbuchophyllia discoidea* showing an early stage in development ×8; (c) *Palaeostrophomena kilbuchoensis,* distorted brachiopod ×1.5; (d) *Liospira striatula*, gastropod ×3; (e) bivalve ×1; (f) *Palaeostrophomena kilbuchoenis* undistorted brachiopod from a mud clast [cf. fig 11. 3(b)] ×1.5.

It is worth recording that in the Leadhills district, though lying outside our area of particular interest, outcrops of similar fossiliferous conglomerates have been recorded over a length of some 10 km. Of these, the localities of Duntercleuch and Snar Water are relatively well exposed; in the former the erosive base of the conglomerate is clearly seen. Other small and badly exposed localities have yielded fossils like those of the

Kilbucho conglomerate, but in none of these Leadhills localities are the fossiliferous mudstones found, only the conglomerates.

How can this unusual sequence be interpreted? We should consider first what the original environment was like, where the now fossilised marine invertebrates lived. They are all shallow-water organisms and the presence of the calcareous alga *Mastopora* in the Leadhills district indicates that the environment in which it lived received some sunlight. It has been established, from palaeocurrent and petrographic studies, that the greywackes of the Kirkcolm formation were derived from a Proterozoic terrane that lay to the north. It is therefore almost certain that the shelly faunas originated from a shallow marine shelf, bordering the Laurentian continent, and facing the Iapetus Ocean to the south. The environment in which they lived may be envisaged as a firm, muddy sea floor, and the diversity and numbers of individuals suggest a high-productivity zone, as might be expected close to a region of coastal upwelling. The massive, poorly sorted conglomerates and the overlying mudstones are consistent with what have been called mass-flow or debris-flow deposits. In this case the larger pieces of rock and fossils would have been supported in a slurry of interstitial water and fine sediment as the unconsolidated sediments slid down into deep water. The cushioning effect thereby provided prevented much abrasion, and ensured that even if the shells did break, the various pieces tended to stay together, hence the crazy-paving effect. The mass flow was sufficiently cohesive to prevent the fossils and larger clasts from settling, but mobile enough to flow quite swiftly. When the flow reached the bottom of the continental slope it would then spread out. The larger rock fragments would have settled first, the smaller ones (i.e. those that formed the mudstones) soon followed. The whole process would be over and done with in a few hours. It is not known if the mass-flow deposits of the Leadhills district and those of Wallace's Cast and Kilbucho were formed in a single contemporaneous event, as it is tempting to believe, or whether there were several smaller flows. Most probably the fauna was derived from the inner part of the shelf, quite close to the shore, and the conglomeratic part of the sequence could well have included shoreline sediments. The distal turbidites of which most of the Kirkcolm formation consists were derived from the outer part of the continental shelf and the continental slope. The mass flow(s) involving these were probably triggered by earthquake shocks one after the other, taking all the materials down to the deep ocean depths. However, the fossiliferous conglomerates and mudstones were derived from the shallow *inner* part of the shelf, and probably the shoreline too.

Clearly some tremendous event must have taken place. If the whole continental shelf had foundered, the deposits would have been much thicker, and chaotic. But the sequence is probably no more than 4 m thick. The most likely interpretation is that a colossal earthquake triggered a great tidal wave or tsunami which devastated the shallow-water shelf, ripping up the sea floor and raking down into the semi-consolidated sediment below. As it retreated, it caused sediments of the shoreline, inner shelf and all their attendant marine invertebrates to slide down into deep water.

The scale of destruction that a tsunami can cause is evident from the tragic events around the Indian Ocean in December 2004, and an indication of the effects on the sea floor is given by the three late Pleistocene and Holocene Storegga submarine slides on the continental shelf off Norway. These are gigantic mass flow deposits, the result of tsunamis, that spread out over the North Sea floor. The oldest was emplaced more than 30,000 years ago and covered an area of 34,000 km^2 whereas the other two were smaller. But collectively the total volume of displaced sediment was 5,580 km^2. The second and third slides occurred between 8000 and 5000 years ago, In the second of these slides in particular are found debris flows like those of Kilbucho, though even if all the occurrences in the Northern Belt formed part of a single slide, it was probably not on such a massive scale.

Faunas of the mass-flow deposits

In the Kilbucho fauna there are at least twenty species of brachiopods, some fifteen species of trilobites, ten of bivalves, and five of gastropods. There are crinoids, strange flat echinoderms, bryozoans and at least one sponge species. The alga *Mastopora* occurs near Leadhills. There is also a remarkable coral, *Kilbuchophyllia*, of which more anon. The faunas are both diverse and abundant (Figs. 11.3, 11.4) and judging by the brachiopods, which include many forms adapted for lying on the sea floor, and also some of the shallow-burrowing bivalves, the sea floor upon or within which they lived must have been fine, tenacious mud.

This fauna is of Upper Ordovician age and compares well with similar faunas from north of the Southern Upland Fault, both at Girvan, in the west, and at Pomeroy in Northern Ireland. The resemblance of the brachiopod fauna, in particular, to that of Pomeroy is very striking. Does this mean that the area from which the Scottish Northern Belt fauna was derived was close to Pomeroy? It may have been. If so, the lateral displacement on the Southern Upland Fault may have been no more than some 300 km. But the Laurentian shelf faunas, living in similar conditions and within the same general temperature zone, may have been much the same, colonising shallow-water habitats that extended far along the coasts. So we simply cannot be sure.

The tale of *Kilbuchophyllia*

We mentioned earlier that a solitary coral, *Kilbuchophyllia*, occurs in the Kilbucho fauna (Fig. 11.3a, b), first described in 1991. Specimens are common at Kilbucho and

Wallace's Cast and in the Leadhills district. In the late 1980s, when a fair number of specimens had been collected, they were sent to Colin Scrutton (Durham University), a noted specialist in Palaeozoic corals, for identification. What he discovered was quite unexpected, and gave a completely new insight into the early history of corals.

There are two main groups of Palaeozoic corals, the Rugosa (Mid-Ordovician–end Permian) and the Tabulata (Lower Cambrian–Permian). They are unrelated to the modern corals (Scleractinia) which began in the Middle Triassic. The rugose corals can be solitary or compound, and very small to large. Now in its simplest form, a coral can be considered as a kind of sea anemone, sitting on a calcareous skeleton which it has secreted from its base. This skeleton may be a discoidal flattish plate, or a curving horn with a concave upper surface in which the living animal resides. The upper surface is traversed by radial walls (septa), which are mirror images of the base of the living coral, which they match exactly so that the animal is firmly held in place. The pattern made by the septa is a primary factor in distinguishing the various groups of corals. In the rugose corals, which had calcite skeletons, there were six primary septa, one being usually shorter than the others. Further generations of septa were emplaced as the coral grew, but these were inserted along four lines only, in a characteristic 'biradial' pattern (Fig. 11.5a). All rugose corals show this kind of symmetry, though in the larger corals it may be obscured by the growth of many subsequent generations of septa. The second group of Palaeozoic corals, the Tabulata, are invariably colonial, and their septa are weakly developed or absent. No tabulates are found at Kilbucho, and they need not concern us here.

Neither of the Palaeozoic groups gave rise to the living scleractinians, which are distinguished by having aragonite skeletons (a different crystalline form of calcium carbonate), and by a quite different pattern of septal insertion from the Rugosa. For the secondary septa are emplaced along six main lines, not four, in between the primary septa, tertiary septa developed in between these, and so on (Fig. 11.5b). It might have been expected that the solitary corals, so common in the Kilbucho fauna, would be rugosans. But no! Colin Scrutton established that this new genus had a pattern of septal insertion identical to that of a scleractinian. This new coral was described as *Kilbuchophyllia discoidea*, with reference to its shape. A further species, *K. clarksoni*, is broadly co-distributed with *K. discoidea*.

So, what is a scleractinian-type coral doing in the Upper Ordovician? Was *Kilbuchophyllia* the ancestor of modern scleractinians (first recorded from the Middle Triassic), which somehow lay in hiding for 200 m.y.? Unlikely. More probably the Rugosa on the one hand, and the scleractinians and *Kilbuchophyllia* on the other, were independently derived from different groups of pre-existing sea anemones. The sea anemones known as the Zoanthinaria, which are biradially symmetrical, gave rise

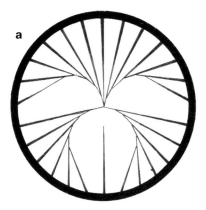 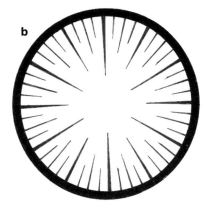

11.5 Patterns of septal insertion in (a) Rugosa (b) Scleractinia and *Kilbuchophyllia*.

to the Rugosa, while the radial Corallimorpha gave rise to *Kilbuchophyllia*, and much later to the scleractinians. The first Rugosa arose in North America, in the late Middle Ordovician. *Kilbuchophyllia* originated just a little later. Yet, whereas the Rugosa thrived to become one of the two dominant Palaeozoic groups, the geographically isolated *Kilbuchophyllia* soon became extinct. Presumably the two groups were competing for the same kind of ecological niches and the Rugosa won. And it was only after both groups of corals became extinct at the end of the Permian that the ecospace they had occupied became vacant. This time it was the corallimorpharian anemones which responded to this potential. Their descendants, the Scleractinia, flourished and evolved into the multifarious corals, solitary and colonial, deep-water and reef-building that we know today.

Had it not been for the great tsunami of some 450 Ma, we would have had no mass-flow deposits, no record of the continental shelf east of Girvan, and certainly no insight into this fascinating episode in the early history of corals, which form so vital a component of the living marine fauna.

CHAPTER 12

The early Devonian: volcanoes and desert storms

By the close of the Silurian and dawn of the Devonian, the Iapetus Ocean had passed into history, although the subducting oceanic plate was still making its presence felt in surface phenomena. Laurentia was now thoroughly joined to Avalonia–Baltica and the combined resultant continent is known by the name Laurussia.

As explained in Chapter 3, 'Old Red Sandstone' is the collective name for the rocks formed out of the erosional debris from the crumbling ruins of the Caledonian mountains. In Scotland, the Old Red Sandstone comprises suites of sedimentary rocks deposited by rivers and lakes, with associated volcanic rocks. Although some of the Old Red Sandstone rocks of the Midland Valley and the Grampian Highlands may date back into the late Silurian from ~415 Ma, it is becoming clear that those of the Borders were probably all deposited within the Devonian Period.

The distinction between the Silurian and Devonian is defined by a dramatic global change in the marine fauna. But in Scotland, marine deposits spanning the Silurian–Devonian break are absent and freshwater fossils are notably scarce. The break from the one Period to the next may well be within the early Old Red Sandstone deposits but is not discernible to the field geologist. The boundaries of different Periods were defined on sharp changes in marine invertebrate faunas. Whereas some Period breaks may be related to major global events, such as the end-Ordovician ice age (Chapter 10) or the terminal Cretaceous meteorite impact, the faunal changes at the Silurian/Devonian boundary are not as marked as with most system boundaries, and it is less easy to account for them. Until comparatively recently the Silurian was thought of as a rather uneventful Period but it is now recognised that it was punctuated both by several extinction events and by sharp, short-lived glaciations. The last of these, in the late Silurian, would have been associated with major sea-level changes that were surely part of the story. However, the precise reasons for the faunal changes at the boundary remain a little mysterious.

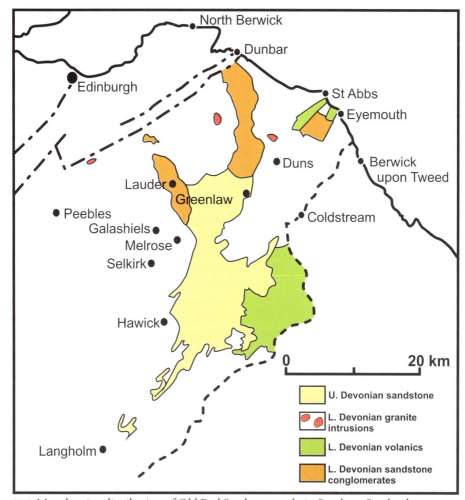

12.1 Map showing distribution of Old Red Sandstone rocks in Southern Scotland.

Palaeomagnetic studies have demonstrated that what is now Scotland lay some 20° to 30° south of the equator at the start of 'Old Red Sandstone times'. It was, however, part of a tectonic plate that was being gently carried northwards, so that, by the end of the Devonian, it had migrated by a further 10° to 20°.

Terrestrial vegetation was in its infancy. The Silurian plants that started to colonise the land were humble organisms, probably restricted to the borders of the lakes and rivers, and the landscape was generally a wilderness of sands, gravels and rocky uplands in a warm climate. Lacking a mantle of vegetation (and consequently soils), run-off following rainstorms was all but instantaneous. Consequently when the rains came, stream erosion was vigorous. Boulders, pebbles and sand were carried down and deposited in the lowlands where subsequent precipitation of minerals between the sedimentary particles cemented them into hard rocks.

The rivers are believed to have been seasonal, replenished after the rains but disappearing in the dry periods, as is commonplace, for example, in many Australian rivers at the present time. Volcanoes were an important part of the landscape in early Old Red Sandstone times, particularly in what are now the western Highlands, the Midland Valley and the Borders Region of the Southern Uplands. Movements along fault planes could abruptly change the topography, damming valleys and causing lakes to develop. The eruption of lava was also capable of blocking river valleys and, together with landslides ('mass-flow deposits') would also have led to lake formation. The nature of some of the Lower Old Red Sandstone deposits leaves no doubt that they were lake-deposited (lacustrine).

Lower Old Red Sandstones of the Borders principally crop out in two belts of poorly cemented and unsorted conglomeratic rocks, with some intercalation of sandstones, to constitute what has been called 'the Great Conglomerate'. Pebbles in this are mainly of greywacke and chert derived from the underlying lower Palaeozoic rocks. The Great Conglomerate, with a total thickness of over 600 m and covering over 200 km^2, formed from flash floods and represents rapidly deposited debris from the newly formed Caledonian mountains. One of these two belts extends south from the vicinity of Dunbar towards the Dirrington Hills near Greenlaw, whereas the other occupies Lauderdale (Fig. 12.1) with another smaller outlier occurring farther north within the Lammermuir Hills. Comparable 'Old Red Sandstones' strata crop out in an area 10–15 km inland from the coast between St Abbs and Eyemouth and are also regarded as part of the 'Great Conglomerate'. Intercalated with the sandstones and conglomerates are thin lime-rich horizons ('cornstones' and 'marls'), the lime coming from the evaporation of waters in which it had formerly been dissolved. The sequence at St Abbs is overlain by a succession of volcanic rocks at least 600 m thick, discussed below.

Early erosion of the Southern Uplands

How much rock has been removed from the Southern Uplands since they formed in late Silurian/early Devonian time remains an open question. It has been estimated that 1 to 1.5 km has been eroded from the Scottish Highlands since they were formed, and in the lower Southern Uplands the figure is probably less. Erosion would have commenced as the Caledonian mountain belt began to rise. The broad valley of Lauderdale is an exhumed fossil feature, filled with Lower Devonian conglomerate. The latter is well exposed along the A68, north of Newton St Boswells, but even better in the Lammermuir Deans in East Lothian (Fig. 12.2), where it is exposed along glacial outwash channels (Chapter 16), with greywacke cobbles up to 30 cm

12.2 ORS conglomerates exposed in the gorge of the Aikengall Water, near Oldhamstocks, SE of Dunbar.

across. The cobbles and pebbles are often imbricated, i.e. lying one on top of another, in inclined sub-parallelism indicative of deposition by flash floods.

Plutonic intrusions

The Ordovician and early Silurian strata of the future southern Scotland had been stacked up and tight-folded as a result of the forces involved in continental collision. As temperatures increased deep in the thickened pile, partial melting of the lower crust took place with the production of magmas that crystallised to granites and related coarse-grained rocks, including diorites and granodiorites. These diverse rock-types can conveniently be collectively grouped as 'plutonic'. This word, intro-duced in Chapter 6, has no compositional connotation but generally implies forma-tion at depths of several kilometres. So, typically, plutonic rocks form where large magma bodies cool slowly to yield coarse-grained rocks. By 'coarse-grained' we mean that individual crystals attain sizes of several millimetres to several centimetres and are thus readily discernible without need for lenses or microscopes. Feldspar is the principal component, commonly present as two species, orthoclase and plagioclase. Quartz becomes important in granodiorites, and especially in the granites, where it constitutes about one-third of the rock.

Whereas the precise nature of granite magma-genesis remains controversial, the crustal melting is likely to have been aided by reduction of pressure as the hot rocks within the Caledonian mountains underwent rapid uplift. At the same time, erosion was stripping sands and gravels from the tops to contribute to the growing pile of Old Red Sandstone sediments. The magmas, being light and buoyant, ascended through the crust to produce the intrusions, with the majority being emplaced between 420 and 400 Ma, when the main phases of deformation had terminated. Some of the larger plutons measure tens of kilometres across, and can be thought of as crudely cylindrical, commonly built up by several pulses of intruding magma. In the con-ventional coloration used on the geological map of Scotland, they are shown as con-spicuous red blots. Although most of these Caledonian plutons were emplaced north of the Highland Boundary Fault, some occur farther south. Among these are some large bodies within the Southern Uplands of Galloway and Dumfries, and some dis-tinctly small examples are present in the Borders. One of these, the Broadlaw intru-sion, penetrates Ordovician rocks close to the Moorfoot escarpment (about half-way between the A703 and the A7). It is roughly 1000 m long by 100 m wide, elongate parallel to the steep bedding of its country-rocks. Two others, cutting the Silurian of the Lammermuir Hills, are the Priestlaw intrusion beside the Whiteadder Reservoir, and the Cockburn Law intrusion, about 3 km WNW of Preston. Although tiny, they may well represent the uppermost portions of large underlying bodies, the bulk of which lies below the present level of erosion. Like the classic, but much greater, Loch Don pluton in Galloway and Dumfries, the Priestlaw and Cockburn Law intru-sions are compositionally zoned, with slightly more primitive (i.e. more calcic and less siliceous) rocks in the earlier, outer parts and less calcic, more siliceous cores that represent a rather younger input of magma. The Broadlaw intrusion may also be comparably zoned with a relatively fast-cooled early facies of 'quartz-diorite' enclosing slightly later and coarser granodiorite. The Broadlaw intrusion displays a well-developed jointing pattern (Fig. 12.3). There is evidence of some baking and hardening ('thermal metamorphism') of the sedimentary rocks surrounding these

12.3 Jointing in the Broadlaw granite as shown in a quarry above the Moorfoot escarpment.

intrusions. Dates for the Priestlaw and Cockburn Law intrusions indicate emplacement at 409 ± 6 Ma. Although the rocks that we now observe at the surface must have crystallised several kilometres down, the presence of granitic pebbles in the lower Old Red Sandstone conglomerates close to Cockburn Law suggests that uplift had been rapid enough, and erosion sufficiently profound, for the pluton to be exposed at the surface relatively soon after its crystallisation.

Fauna and flora

Terrestrial vegetation evolved fast during the lower Devonian although it was still probably confined to the wetlands. In the Borders, plant remains from this time are extremely rare, but we can obtain some conception of what it was like from the world-famous Rhynie Chert in Aberdeenshire – the subject of extensive research for nearly a century. This is effectively a fossilised peatbog, where plants and animals have been preserved in three dimensions by silicifying water from hot springs and geysers. The plants were relatively small, none being more than 50 cm in height. Most belong to an extinct group, the rhyniophytes, in which there were no leaves, but there are also early representatives of club-mosses (Lycopodiaceae); these latter have 'scale leaves' as in a monkey-puzzle tree.

In the cooler pools small crustaceans (*Lepidocaris*) swam, and millipedes and spider-like trigonotarbids lived amongst the plants. Lesions on the plant stems testify to early arthropods eating the plants, and presaging the insect/plant relationships so common today. How representative the Rhynie flora is of Lower Devonian vegetation remains an open question, but it seems reasonable to conclude that the plant and animal communities living in the Borders were not dissimilar to those of Rhynie.

The jawless fish of the Silurian persisted into the Lower Devonian, including such forms as *Cephalaspis* but there were also the earliest fish with jaws, including the 'spiny sharks' or acanthodians, in which the jaws, rather like false teeth, were not actually attached to the skull. This was also the heyday of the giant water-scorpions (eurypterids), which inhabited fresh to brackish water. *Pterygotus* was an active predator up to 2 m in length, with pincer-like claws.

Old Red Sandstone volcanoes in the Borders

Volcanic rocks of 'Old Red Sandstone age' occur in two parts of the Borders: a) near Eyemouth and St Abbs, and b) some 25 km further south where they form the Cheviot massif.

The Eyemouth Volcanic Sequence

The Eyemouth lavas appear to have erupted prior to 400 Ma and may predate the Cheviot eruptions by several million years. The volcanic rocks exposed along the coast between St Abbs and Eyemouth (Fig. 12.4) are inferred to have been erupted from a group of volcanoes strung out along a line trending NW–SE. The lavas and associated fragmental deposits are seen at their best along the coast between Pettico Wick and White Heugh (Fig. 12.5). The sequence generally dips towards the SE at 30–40° but is broken up by a number of NE–SE-trending faults that mainly downthrow towards the south. This succession lies to the seaward side of the NW–SE-trending St Abbs Head Fault and the vertical fault plane itself can be seen on the cliff-top at Hardencarrs Heugh (NT 9176 6803) as a 2.5 m-wide zone of fractured rock ('fault breccia'). The fault has displaced the volcanic succession downwards relative to the rocks on the landward side, probably through several hundred metres. The rocks to the west comprise strongly folded Silurian greywackes, unconformably overlain by basal Old Red Sandstone conglomerates on Bell Hill. This conglomerate sequence, some 120 m thick, is composed of pebbles and cobbles derived from the Silurian 'basement', and the absence of volcanic pebbles indicates that volcanism had not yet started when the conglomerate was forming.

From the sparse evidence available, one can try to construct an image of the contemporary landscape. A chain of andesite cones is envisaged, possibly rising to

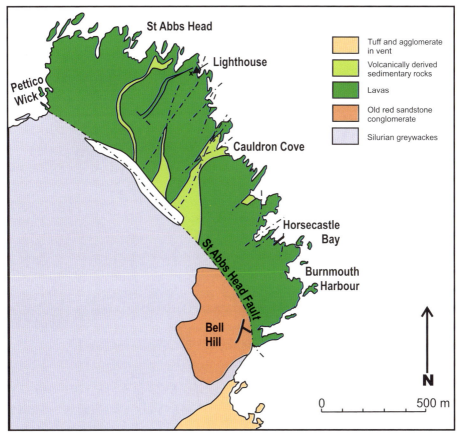

12.4 Geological sketch-map of the volcanic rocks near Eyemouth – St Abbs.

heights of a few hundred metres. These may have been sited in a wet environment, perhaps within a lake retained in a subsiding, fault-bounded, rift basin. Footprints of an eurypterid, *Pterygotus* (described above) have been described from the sediments, implying relatively tranquil conditions and intervals when there were wet sandy lake shores. Infrequent eruptions, separated by several hundreds of years, would have seen emission of slow stodgy, viscous lavas that flowed, at most, only a few kilometres from the volcano craters. Incandescent lava flowed within clinkery banks of already cooled lava, much as a river may flow between raised levées. Blocks of largely crystallised lava would have spilled down from the advancing front of the flow as the hot and mobile interior pulsed forwards. The lavas from any particular eruption ranged in thickness from a few metres to ~50 m. The separation of volcanic gases from the lava produced abundant of bubbles (vesicles) within the molten lava, with sizes up to 2 cm. The slow movement of the lavas, continuing until after the gases stopped being released, twisted and stretched the vesicles into contorted shapes. Then, long after the

12.5 Lavas and pyroclastic layers in the Eyemouth Group. The abrupt changes in dip are due to faulting.

eruptions had ceased, percolation of warm aqueous solutions through the permeable and porous lavas caused the precipitation of secondary minerals, filling the vesicular cavities to produce what are referred to as amygdales.

Warmth, infrequent rain showers and a sufficiency of time typically reduced the jagged lava surfaces to a clay-rich material, reddened with iron oxides. These deeply weathered red upper surfaces, up to 2 m thick, testify to the long time lapses between one eruption and the next. Thin sedimentary horizons separating some of the volcanic units are largely made up of sand-grains washed down from the eroding volcanic outcrops.

The flanks of the volcanoes were sufficiently steep for masses of fragments to break loose now and again, to slide downhill as mud-flows (or 'mass-flow' deposits) (Fig. 12.6). Such flows would have been promoted when rain-sodden masses of sand and rock fragments became unstable and produced what are known as 'lahar' deposits. In these, the fragments are present as a chaotic jumbled mass, lacking any sorting with respect to size, shape or composition – and hence lacking any discernible stratification. The waterlogging required to produce these gravitationally unstable masses could have come about from intense desert rainstorms or from rain resulting from steam-laden clouds attending eruptions. Lahars are common in volcanic regions and are one of the most frequently encountered hazards for those living on or near active

12.6 A mass-flow deposit or 'lahar' – unbedded, jumbled mass of volcanic fragments in fine-grained matrix. Coast at Eyemouth.

volcanocs. Lahars are a phenomenon common on the volcanoes of Java, and the word is a Javanese term.

Lamprophyre dykes

Another suite of magmatic intrusions plays a role in our story. Unlike the coarse-grained, often sub-cylindrical plutonic intrusions, these occur as a swarm of thin (typically less than 3 m wide), parallel-sided and vertical sheets (Fig. 12.7). Such intrusions are known as dykes and are typically emplaced in crust experiencing stretching. The dykes in question constitute a swarm that runs parallel to the 'grain' of the deformed lower Palaeozoic strata (i.e. ENE–WSW), essentially parallel to the Iapetus Suture along which the ocean closed.

The rocks forming the dykes are called lamprophyres. The word implies a 'phyric' rock, that is, an igneous rock with a fine-grained matrix but containing larger, earlier formed crystals ('phenocrysts'). The first syllables come from the Greek word 'lampros' meaning 'shiny'. In these lamprophyres the shiny phenocrysts are crystals of the dark mica, biotite, which can constitute between 20 and 60% of the rock. Chemically, the abundance of biotite makes them unusual and implies a high content of potassium. This, together with other geochemical signatures, categorises them as rock-types closely associated with subduction.

12.7 The 'Horsecastle Rock', a lamprophyre dyke cutting volcanic rocks of the Eyemouth Group, near Coldingham. The dyke rock is more resistant to weathering than its volcanic wall-rocks and consequently is left outstanding as a prominent wall.

These potassic lamprophyre dykes form a swarm that is only about 10 km broad but which is traceable over ~ 400 km, from just south of Belfast (the Ards Peninsula) ENE across the Southern Uplands to St.Abbs on the Berwickshire coast. Although closely associated in time and space with the plutonic intrusions, a date of 400 ± 9 Ma on one of them suggests that the dykes came in several million years later.

An idea outlined by Paul Shand and his co-workers is that melts rising from the subducting Iapetus Oceanic lithospheric slab affected the overlying mantle wedge (cf. Fig. 4.6) by adding water, potassium and other components to it and, in so doing, lowered the temperature at which it starts to melt. Although only a very small proportion of the mantle rock would have melted to form the lamprophyre magma, it was this melt that contained the bulk of the water, potassium and other distinctive elements. Thus the potassic lamprophyres are inferred to be the frozen magmas that resulted from small amounts of melting of the chemically modified mantle.

Deposition of the 'Great Conglomerate' is believed to have occurred between 410 and 400 Ma and thus probably overlapped the episodes in which the plutons, the St Abbs–Eyemouth volcanic sequence and the lamprophyre dyke swarm were generated.

The Avalonian crust was underthrust beneath the accretionary pile of Lower Palaeozoic sediments (Chapter 6) and, according to Shand and his co-workers, the lamprophyre dyke swarm may actually demarcate the northern limit of this under-thrusting. And, as also stated earlier, although the Avalonian continental crust would have been too light to have followed the dense Iapetus lithosphere far down into the mantle, it may have been dragged some way down the subduction zone before coming to a halt (cf. Fig. 6.6).

The Cheviot volcano

The Cheviot is the most commanding physical feature in the Borders, rising to a height of 815 m. It is composed of igneous rocks, straddling the Scotland–England border, probably somewhat to the north of the Iapetus suture. The Cheviot lavas outcrop over a crudely circular area, with a diameter of some 60 km (Fig. 12.8). Although only about one-third of the outcrop lies in Scotland, we give here an outline description of the entire massif. An age for the Cheviot igneous rocks, established in 1972 as in the range 400–380 Ma, was subsequently refined as approximately 396 Ma, making the Cheviot magmatism some ten million years younger than the majority of the Old Red Sandstone-age volcanoes that erupted in the Midland Valley and Grampian Highland terranes and also younger than the Eyemouth–St. Abbs volcanism.

Whereas in the early stages of volcanism a number of separate vents may have been involved, the whole Cheviot assemblage probably represents the eroded remnants of a great 'central-type volcano', very much larger than the Eyemouth and St Abbs volcanoes. A large part of the volcanic sequence doubtless lies buried beneath the younger Carboniferous sedimentary rocks on the eastern side of the Cheviot. Outcrop is generally poor and the rocks have been comparatively neglected by geologists over the last thirty years. Much of the discussion presented here is based on descriptions by D.A. Robson and J.G. Mitchell.

Topographically the Cheviot presents a deeply dissected plateau divisible into an inner plateau (600–800 m.) and an outer plateau lying at a height of ~450–500 m. The high ground has a roughly radial drainage pattern with some deeply entrenched V-shaped valleys, some of which (e.g. the College and Harthope valleys) were controlled by NNE- and NE-trending faults (Fig. 12.8). The Gyle–Harthope and Thirl Moor faults trend NE–SW and represent rejuvenation of the 'grain' of the underlying Southern Uplands originally imposed during the Caledonian Orogeny, and all were ultimately inherited from the lithospheric weaknesses induced at the time of Iapetus Ocean closure.

The Cheviot lavas are sandwiched between previously deposited sandstones and very similar younger rocks. Thus the volcanism was merely an interlude in the evolution of the Old Red Sandstone landscape in which the processes of deposition and erosion otherwise continued unabated. Literally, the Cheviot activity appears to

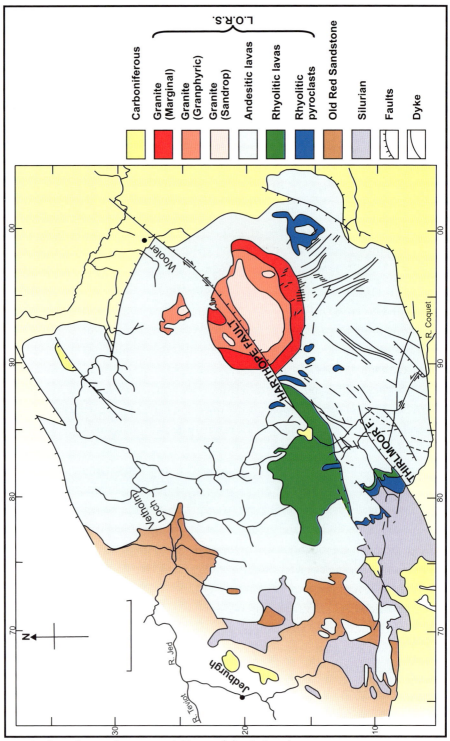

12.8 Geological map of area around Cheviot (after Robson).

12.9 Bedded tuff in the Gaisty Burn, SW Cheviot.

have started with a series of bangs, since the earliest volcanic products are pyroclastic rocks signifying highly explosive eruptions. Robson concluded that these could have involved numerous volcanic centres, building up cones of fragmental debris. Whilst these would have experienced severe contemporaneous erosion, some of the pyroclastic rocks were preserved beneath subsequent lava flows, and their remains can be found in faulted blocks on the western side of the massif. The best outcrop occurs in the Gaisty Burn (south of Gaisty Law) where a succession some 70 m thick occurs (Fig. 12.9) ranging from fine-grained tuffs to very coarse 'agglomerate', containing blocks up to 1 m across. The early pyroclastic rocks are overlain by a sequence of rhyolite lavas. The large size of the blocks in the agglomerates indicates that the outcrop probably lies not more than a few hundred metres from the site of the eruptive vent.

Thus the first magmas to reach near-surface levels were rhyolitic. Such magmas are not only rich in silica but are also typified by high contents of dissolved water and other potentially volatile components. The reduction in pressure as the magmas near the surface allows their dissolved gases to escape violently, as outlined in Chapter 10. Although the high silica content of rhyolitic magma confers relatively high viscosity, the viscosity increases still more as the water comes out of solution, so that the gases escape only with great difficulty from the fast-congealing magma. Gas pockets build up pressure until they are explosively discharged, blasting shards of the semi-solid rhyolite, together with fragments of already solidified

12.10 Terraced topography formed by andesite lavas. Woden Hill, west of Cheviot.

rock, high into the air. Any remaining rhyolite magma, having lost the greater part of its volatile components, could then be extruded as slow-moving lavas, oozing through the craters created by preceding explosions. Such lavas rarely flow more than a few hundred metres from their source. A later group of rhyolite lavas and interbedded pyroclastic beds gives bare crumbly slopes and pinkish gashes on the north-west face of Beefstand Hill.

Most of the Cheviot is, however, composed of the rather less silica-rich (and correspondingly more fluid) andesite lavas. The rubbly and weathered tops of the lavas weather more rapidly than the interiors and consequently lava profiles on the hillsides may present a terraced or stepped appearance. The tops erode back to form slack ground, whereas the tougher interiors form steeper escarpments (Fig. 12.10). Slow cooling of thicker lavas can result in contraction cracks (joints) growing roughly perpendicular to top and base. While the andesites are generally very fine-grained, they contain a proportion of larger crystals ('phenocrysts') up to a few millimetres long. These typically comprise three minerals, plagioclase (feldspar), augite and hypersthene. The phenocrysts are those that started crystal growth before eruption but became encapsulated in the fine-grained matrix produced by fast crystallisation of the melt as it rapidly lost heat and volatiles during eruption. Indeed, the ensuing congelation was commonly so rapid that crystallisation was prevented and instead, the melt merely super-cooled to a natural glass. Thus some of the Cheviot lavas are

12.11 Columnar jointing in cliff-face south of Kirk Yetholm.

andesites in which the phenocryst minerals are set in a glassy matrix. On the north-facing hillside above Kirk Yetholm is a conspicuous cliff, some 6 m high, composed of such largely glassy material (Fig. 12.11). Probably this represents a thick unit, extruded as a prominent lava dome or blister that cooled yielding well-developed columnar joints as it contracted. The rock-type is most easily examined in the Kirk Yetholm church itself, which has been built from this material.

Whereas the Cheviot is predominantly composed of lavas and pyroclastic rocks, intrusive rocks occur near the summit, forming a roughly circular outcrop, some 10 x 8 km across. Although the intrusive complex is eccentrically offset towards the south-eastern side of the massif, it may originally have lain centrally, beneath the volcano summit: if so, the volcano spread further to the east than indicated by the present lava outcrops, with a significant part being hidden beneath the Carboniferous cover.

The intrusions resulted from the ascent and emplacement of large magma bodies that attained high levels in the volcanic superstructure late in the history of the volcano. Their slow cooling resulted in the production of rocks more coarsely grained than the lavas. The crystals are generally several millimetres across whilst in the lavas the grain-size is typically 1 mm or less. The first of these intrusions was a diorite grading to granodiorite. (A diorite is compositionally the equivalent of an andesite, while a granodiorite equates compositionally to somewhat more siliceous (dacite) magma.)

After emplacement of the diorite–granodiorite there was further intrusion of more siliceous magma and, because of the higher silica content, the crystalline products contain significant amounts of quartz and they are classed as granite rather than diorite or granodiorite. The magma from which the granite crystallised would have been compositionally related to the rhyolites. The various intrusive types are distributed in a roughly concentric fashion, with the diorite/granodiorite forming an incomplete outer ring and the granite (or granites) lying within it. So there is a crude zonal sequence comparable to those described above for some of the earlier plutonic intrusions.

We suggest that emplacement of the intrusive complex was probably accompanied by successive collapses of the volcano summit leading to formation of large steep-sided pits known as calderas. Caldera collapse is commonly brought about by the violent evacuation of large volumes of magma from high-level magma chambers within the volcano. Silica-rich magmas such as dacite and rhyolite are particularly prone to erupt with emission of great clouds of hot gas containing high concentrations of molten and solid particles. Such dense gas- and particle-rich clouds tend to be ground-hugging and move fast, much in the manner of snow-avalanches, ultimately coming to rest and collapsing as gas and particles separate out. These volcanic avalanches are called pyroclastic flows, and it may be supposed that such eruptions attended the latest stages of the volcano. The dioritic and granitic intrusions doubtless crystallised beneath a thick cover – probably largely composed of extrusive materials that could well have been largely composed of pyroclasts produced during the explosive degassing of earlier bodies of magma. Figure 12.12 shows how the volcano might have evolved.

The final magmatic stage at Cheviot was marked by intrusion of a plethora of dykes that cut both the lavas and the main intrusive centre. Four different categories of dykes have been recognised, each of them relatively silica-rich. That the dykes intersect the dioritic/granitic intrusions indicates that, even after these had been emplaced and slowly cooled, relatively silicic magmas were *still* available at depth to rise into propagating fissures.

The magmas responsible for the Cheviot complex probably took advantage of a weak zone in the lithosphere. As indicated in Chapter 4, basaltic magmas are by far the most abundant compositional type produced by terrestrial volcanoes, and it is of interest that of all the variety of magma types that went to make up the Cheviot, basalt is strikingly absent. The andesitic and subordinate dacitic and rhyolitic magmas that went to build the structure are all types for which an environment above a subduction zone could be inferred. This is equally true for the Eyemouth lavas, which suggest that all of these magmas were related to a subducting slab. But this poses

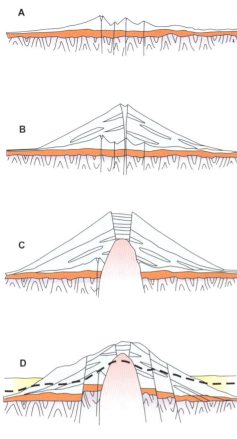

12.12 Sketch diagram of possible Cheviot volcano evolution. Hypothetical cross-sections of the evolving Cheviot volcano. **A.** Explosive eruption of rhyolite pyroclasts and lavas. Possibly from several vents. **B.** Construction of a volcano composed dominantly of andesite lavas, pyroclasts and lahar deposits. **C.** Ascent of more silicic magma to form large magma chambers within the volcano superstructure. Surface eruptions (largely pyroclastic) of andesitic and dacitic composition, were associated with caldera collapse. Magmas retained at depth crystallised to diorites and granodiorites. **D.** Erosion persisted during and after the growth of the volcano. Block-faulting, in Middle Old Red Sandstone times affected the extinct volcano. Younger Devonian and Carboniferous sedimentary deposits partially covered the volcanic rocks. The thick dashed line represents the current erosion level.

12.13 Impression of the Cheviot volcano during its andesitic stages. The foreground depicts wetlands with early Devonian flora.

a dilemma: we see the results of abundant early Old Red Sandstone volcanism in the Midland Valley and Highlands, now forming the Pentland, Ochil and Sidlaw Hills as well as those of the Lorne Plateau in Argyll. However, these all date from 415–410 Ma, i.e. some 10 to 20 m.y. earlier than the St Abbs–Eyemouth and Cheviot volcanism. The magmas responsible for these older volcanoes are inferred to have arisen from melting on top of the Iapetus Oceanic lithosphere, the slab still continuing descent after the collision and closure of the ocean had occurred. The ages of the St Abbs–Eyemouth and Cheviot volcanoes obviates the possibility that these were formed from the same subducting slab. It invites speculation that mantle melting may have taken place above one or more subordinate pieces of oceanic lithosphere. In other words, the subduction may not have involved a *single* continuous lithospheric slab plunging beneath Laurentia but may have broken into smaller units.

While the area and height of the Cheviot volcano remain matters for speculation, Robson's mapping indicates the total thickness for the extant lavas to be at least 500 m, whereas Mitchell and co-workers suggested that the andesites themselves could possibly be 2000 m thick. Robson also noted that the intrusions of the summit area could have crystallised beneath an extrusive cover of perhaps 1 km thickness. With a base diameter of up to 60 km and so large a thickness of extrusive products, the volcano may well have risen up to 3 km. Mt. Etna on Sicily could be a crude analogue so far as size is concerned – although it is very different compositionally!

CHAPTER 13

Late Devonian landscapes and sedimentation

Younger Old Red Sandstone rocks crop out from Duns and Bonchester Bridge, south-westwards as a narrow, discontinuous belt, through Langholm towards the Solway Firth (Fig. 12.1). They consist of sandstones, red pebbly sandstones and conglomerates deposited by braided, meandering rivers. Some came from the hills of what is now Galloway, draining towards the north-east, but those of the Pease Bay, Kelso and Greenlaw areas of the Borders were deposited by rivers flowing south-westwards. It is deduced that 'southern Scotland' lay in an area of inland drainage with seasonal lakes developing in a generally arid climate. Lake Eyre in Australia or the Aral Sea in central Asia are approximate modern analogues.

In the larger eastern outcrops, the rocks are largely soft, dark red-brown and rather fine-grained sandstones, mudstones and marls (Fig. 13.1), together with 'cornstones'. These were formed when occasional rains percolated down through the porous sediments, dissolving carbonates as they did so and, in the dry seasons when evaporation was high, the waters were drawn back up by capillary action. As they evaporated, the carbonates were precipitated, forming generally irregular hard clumps to which the name 'cornstones' is given. Figure 13.2 shows the appearance of such cornstones in strata close to the Devonian–Carboniferous boundary, near Burnmouth on the Berwickshire coast.

Plants and animals

Some 15 m.y. before the end of the Devonian a series of environmental crises took place, spread over three million years or more. These crises severely affected the marine habitat and are collectively regarded as one of the five great mass extinctions that revolutionised the history of life on Earth. Worst affected were the tropical

13.1 Upper Old Red Sandstones near Denholm.

reef ecosystems and warm-water shallow marine communities. However, the high-latitude ecosystems and deep-water communities suffered less severely and the terrestrial ecosystems were largely unscathed. There were probably a number of contributory factors involved, including global cooling. Yet sea-level stood high for all this time and there is no evidence of glaciation during the late Devonian. Ocean current systems, however, were affected, and one likely scenario envisages an 'oceanic overturn' whereby stagnant sea-floor mud became spread through the upper waters of the sea, poisoning them and leading to faunal extinctions on more than one occasion.

However dramatic were the extinction events in the marine realm, there is no trace of them in the Upper Old Red Sandstone strata in the Borders, where fossils are very scarce. But nonetheless, extremely important events were talking place on shore, since it was during this period that the invasion of land by plants became well established. We have seen that in the Lower Devonian, plants were small and lived close to water, but by the Middle Devonian the first real forests were flourishing, and diversity was increasing impressively. The Late Devonian saw the first seed-bearing plants but, at around the same time as the major marine extinctions were occurring, there was a corresponding drop in the diversity of land plants. Why and how this happened, again we do not know. In favoured areas there were forests of quite large trees dominated by club-mosses, precursors of the giant lycopods of the Carboniferous. Ferns, creepers and tree-ferns thrived, and some other plants with fine-sounding names, *Psilophyton*

13.2 Cornstones in Lower Carboniferous (Upper Old Red Sandstone) sandstones at Burnmouth.

and *Zosterophyllum*, survived from the Lower Devonian. In most respects the vegetation of the Upper Devonian was not dissimilar to that of the Carboniferous, as typified by the coal-swamp plants, but less diverse, sparser and with smaller plants.

In Upper Devonian times, fish had become abundant in fresh waters as well as in the sea. Yet whereas the marine forms suffered severely during the late Devonian extinctions, the freshwater fishes did rather better. All jawless fish disappear from the record although the lines that led to modern lampreys and hagfish must have survived somewhere. Freshwater placoderms and acanthodians persisted, although reduced in diversity. As we mentioned in our former book, many specimens of the placoderm *Bothriolepis* were found crowded together at a locality south of Pease Bay, presumably trapped in a shrinking pond during dry weather, where they died of suffocation. (This is one of the few places, to our knowledge, that Upper Devonian fossils occur in the Borders.) Most important, however, are the bony fishes (Osteichthyes), for amongst these are several groups of lungfish in which part of the gut modified into a lung. The ephemeral nature of lakes and rives in the hot dry environment has been referred to above, and consequently these air-breathers were well adapted for survival. It was from one group, the lobe-fin fishes (Crossopterygians), that the first tetrapods evolved, the four-legged amphibians which were the first vertebrate colonisers of the land. The transformation of the small bones in the lobe-like fins of these fish, especially Eusthenopteron, into the hand and leg bones of the early amphibians has been meticulously documented over the years, and it is now known that the earliest amphibians originated just before the late Devonian mass extinctions.

Arthropods of various kinds successfully invaded the land, as well as freshwater environments. Millipedes and early 'insects' colonised the terrestrial environment and in the fresh waters, the giant water-scorpions (eurypterids) continued to thrive, though contemporaneous marine forms were affected by the late Devonian crises. Clearly, life on land and in fresh water was flourishing in most parts of the Borders during Upper Devonian times.

CHAPTER 14

Rivers, lakes and shallow seas of the sub-tropical 'Borders'

A broad swathe of country around the northern, eastern, and southern sides of the Cheviot is underlain by Lower Carboniferous strata. On the Scottish side of the border we see these, along and close to the coast of East Lothian and Berwickshire between Dunbar and Cockburnspath, and south of Eyemouth, between Burnmouth and Berwick-upon-Tweed. The Lower Carboniferous also crops out between Duns and Kelso and as a narrow belt trending NE–SW from Carter Fell towards Canonbie and Langholm (Fig. 14.1).

The Devonian, which ended at 359.2 ± 2.5 Ma, was succeeded by the Carboniferous Period, which went on until 299 ± 0.8 Ma. Had one been around at the close of the Devonian, nothing obvious would seem to have occurred in the Borders: one day was followed by another as seamlessly as the 31st July is by the 1st August. Globally, however, the break between the one Period and the next was marked by a widespread extinction of marine organisms that took place in several pulses, the strongest actually being before the end of the Devonian. The causes of the extinction are debatable, one favoured option being an immense overturn of stagnant bottom water. The movement of tectonic plates, however, controlled by the relentless convective overturn in the mantle, was certainly changing the geographic disposition and shapes of the continents. A knock-on effect of this would have been continuous changes in the pattern of ocean currents, and the rise of anoxic and lethal bottom water into the upper waters of the sea could well have been one of the consequences.

That part of the tectonic plate carrying the Borders was being shifted progressively northwards, from a position around 25° south of the equator in the early Devonian to about 15° degrees south in the early Carboniferous, so roughly equivalent, say, to Zambia or northern Mozambique at the present time. Not only was the latitude

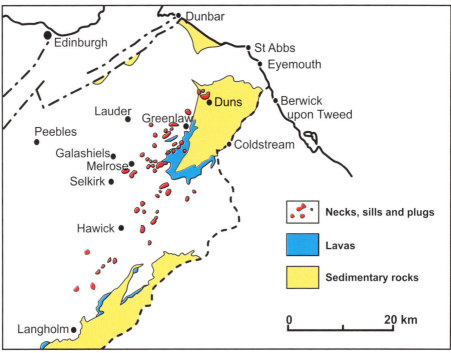

14.1 Distribution of lower Carboniferous sedimentary and igneous rocks in the Borders.

changing, but the climate also. At the start of the Carboniferous the climate in the Borders was warm and relatively dry. As 'Scotland' migrated northwards towards the equator, the sedimentological evidence shows the rainfall increased to a maximum before declining to give extreme desert conditions at the close of the Carboniferous Period. In the Borders, however, we have only a record of conditions in the early Carboniferous.

At this time there lay to the north the Lower Palaeozoic mountains which now form the Southern Uplands, while to the south was the larger Northumberland trough which was continually subsiding during Lower Carboniferous times and receiving mainly marine sediments (Fig. 14.2). Between the northerly mountains and the sea lay an embayment to the north of the Northumberland trough, known as the Tweed Basin. The basic story which the Carboniferous rocks of the Tweed basin have to tell is of desert-fluviatile sandstones (Fig. 14.3) like those of the under-lying Upper Old Red Sandstone being replaced by semi-marine sandstones and shales and then eventually by fully marine sedimentation as sea-levels rose and shallow seas encroached across the lowlands. The low-lying landscape was punctuated by several dozen volcanoes that are the subject of the next chapter. The sedimentary sequence is not quite the same in the more northerly part of the area, however, as it is in the south, for the swamplands were missing in the north.

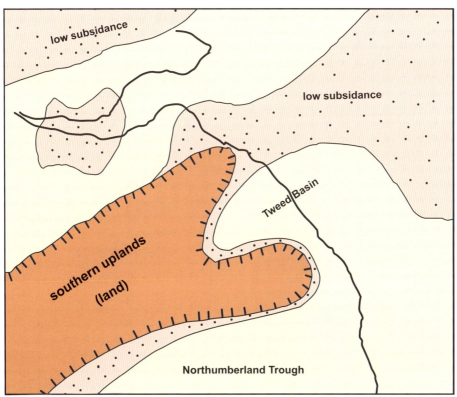

14.2 Palaeogeographical map of southern Scotland during the early Carboniferous (after H. Francis). The modern coastline is also indicated.

14.3 Horizon bedding in pale-coloured deltaic Lower Carboniferous sandstones. Carbonised traces of plant fossils are abundant in these strata. Quarry at Swinton, *c.*8 km SE of Duns.

Burnmouth

The shore section near the fishing village of Burnmouth displays some spectacular geology, and is well worth visiting (Fig. 14.4). The lowest beds, well exposed north of the harbour, are folded Silurian greywackes, shot through with Siluro-Devonian

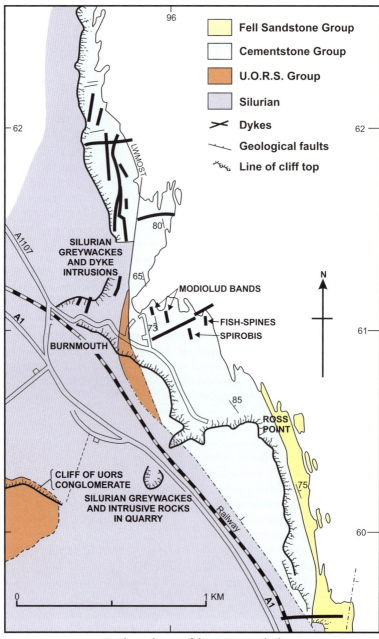

14.4 Geological map of the Burnmouth shore.

age dykes. Elsewhere in the section are younger quartz-dolerite dykes of probable late Carboniferous age. The Silurian is faulted against younger rocks of late Devonian and early Carboniferous age, all of which have been steeply rotated by faulting.

The (late Devonian) Upper Old Red Sandstone succession is fluviatile in origin. This mainly consists of reddish-brown sandstones with pebbly horizons, passing gradationally upwards into the Carboniferous Cementstone Group. This latter consists largely of unfossiliferous mudstones, siltstones, and thin sandstones, but also with the 'cementstones' that give the group its name. These are beds or nodules of pale-coloured, fine-grained sandy or muddy dolomite, resembling hardened cement. The whole assemblage containing the 'cementstones' is known as the 'Cementstone facies', of which some 500 m is exposed at Burnmouth. Some 30% of the succession consists of thick cross-bedded sandstones, fining upwards with sharp erosive bases (Fig. 14.5) and the remainder is mainly sandy mudstones, thin sandstones and cementstones that tend to occur in cycles (cementstone–mudstone–sandstone–mudstone–cementstone). Fossils are scarce (mainly bivalves and worm tubes), but those that do occur are all non-marine, and most of the enclosing sediments were deposited in rivers and lakes. The commonest are bivalves, but there are also calcified worm tubes and fish spines.

We can envisage a broad muddy floodplain, studded with shallow lakes, stretching southwards from the high Silurian ground to the sea which lay to the south (Fig. 14.6). The rivers frequently broke their banks, and floods of sandy material spilled out onto the muddy floodplain; these deposits are interspersed with the muds and siltstones that make up the background floodplain sediment. The origin of the cementstones themselves is harder to interpret. Those that are bedded sometimes contain laminations, desiccation cracks and other sedimentary structures, together with ostracodes and bivalves. They are probably original rather than diagenetic, and much of the carbonate may have been derived from degraded algal debris. Perhaps the most likely scenario is that they formed in temporary lakes as algal carbonate mud. The nodular cementstones, on the other hand, may have been diagenetic in origin. The lakes formed when the rivers became temporarily blocked, for example by the uplift of fault scarps or by a lava flow from one of the volcanoes. Plants grew round the lakes in some profusion and probably covered parts of the floodplain too. The floodplain extended southwards into muddy coastal flats, which were subject to frequent marine inundation when strong winds blew onshore.

The Fell Sandstone Group lies directly above the Cementstone Group; it consists of pale-coloured, yellow, pink or white sandstone and is usually cross-bedded and sometimes convoluted. This group of sandstones is up to 400 m thick, forms the extensive hill country west of Alnwick and represents an extensive, long-lived

14.5 Lower Carboniferous strata at Burnmouth, Berwickshire tipped vertical by faulting. Sandstones are shown exhibiting cross-bedding ('current-bedding'), 'younging' from right to left. The cross-bedding present in the prominent unit in the centre of the picture will have been steepened by movement ('soft-sediment deformation') when the sands were still wet and readily deformable before becoming lithified into hard rock.

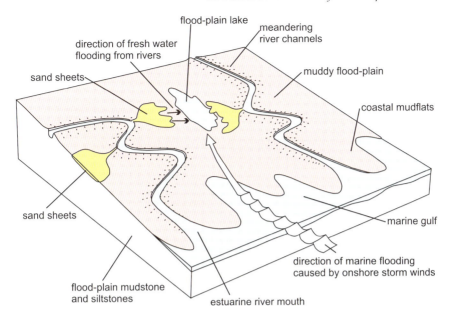

14.6 Palaeoenvironmental reconstruction of the Lower Carboniferous Cementstone environment (after R. Anderton).

delta deposited from braided rivers and is evidently part of the same river system which flowed from the north-east. When this delta finally ceased to function, and at a time when rainfall had substantially increased, a further kind of sedimentary succession was deposited. These sediments, The Scremerston Coal Group, consists of cross-bedded sandstones of deltaic origin, often cutting into the sandstone below, but there is an extensive marine limestone, replete with colonial corals, brachiopods and other fossils. The marine incursion did not last very long; there was a return to deltaic sedimentation and growth of large trees, typical of coal forests, on the delta surface. There are some twelve coal seams, irregularly spaced within this Group, and whereas none of these is more than 25 cm thick, they were extensively worked at one time, and the relics of ancient mines are visible in some places along the shore. Such coal seams suggest a largely green and pleasant landscape; plants had progressed not only in space and time but also in diversity. In these well-watered lowland swamps the flora was dominated by gigantic club-mosses (lycopsids), like the well-known *Lepidodendron*, horsetails such as *Calamites,* and various kinds of ferns (sphenopsids), progenitors of the late Carboniferous, coal-forming 'swamp-forests'. And the lakes and waterways were habitats for a wide variety of fish of increasingly modern type, replacing the more heavily armoured fish that characterised the Devonian. Among the most conspicuous of the invertebrates were the eurypterids, of other genera than we have seen in the Devonian, and which were the top predators of the time.

One may wonder why the Carboniferous beds in the vicinity of Burnmouth are so strongly tilted and, in places, vertical. This is because they belong to the Berwick Monocline, in which the landward beds are very steep, whereas the seaward beds dip gently. This feature was initiated during a late Carboniferous orogeny that resulted from continental collision far to the south-west, which formed the intensely folded, faulted and metamorphosed structural belt of Devon and Cornwall. Yet its effects were felt further north, and include the uplift of the Pennines and the updoming of the Lake District. The Berwick Monocline is perhaps the furthest north major fold resulting from this immensely powerful episode in Earth history.

Limestones are also beautifully exposed on the shore east of Berwick, strikingly formed into folds which have been planated by the sea. There are two major faults also. These, like the Berwick Monocline, noted earlier, are all Hercynian structures, related to major earth movements far to the south.

Foulden

We have noted above that the Cemenstones were evidently deposited on a low lying floodplain, stretching southwards between the lower Palaeozoic mountains to the north and the sea. This floodplain was crossed by small meandering rivers, passing into coastal mud flats bordering the deep and fully marine Northumberland trough. Some semi-permanent lakes were scattered on the surface of this floodplain, teeming with abundant life (Fig. 14.6). One of these lakes is represented today by sediments exposed in a ravine at Foulden Farm, some 8 km west of Berwick, and just inside the Scottish border.

A young local collector, Thomas Ovens, first found fossil fish at this locality in about 1910, and amassed about 150 specimens. When he died in 1912, at the age of nineteen, his parents sent all the specimens to the Natural History Museum in London. Some time later they were described and illustrated by the foremost authorities of the day. In the early 1980s the site was re-located, and properly excavated under the auspices of the National Museums of Scotland, with the redoubtable collector Stan Wood carrying out the bulk of the work. The succession at Foulden lies about 100 m above the base of the Cementstone Group, and some 30 m of sediment are exposed. These are mainly uninspiring silts and sands, and largely unfossiliferous, but towards the top of the exposed sequence is the Fish Bed, which consists mainly of finely laminated siltstones with very thin cementstone layers.

The lake, during its lifetime, was shallow and only a few metres deep. Since none of the fossils are of unequivocally of marine origin, the lake must have been brackish or freshwater. The beginning of lake sedimentation, following standard floodplain deposits, is marked by a thin shell-bed, rich in the bivalve *Myalina*, which formed a solid substrate. Above this there is a plant bed, some 40 cm thick, and then the fish bed proper, some 1.1 m thick. The fauna of the fish bed is dominated by swimming animals (fish and crustaceans –Figs. 14.7, 14.8) and the lake floor, though oxygenated, was normally too soft to support a benthic fauna until the last stages of its existence.

14.7 Fish and arthropods from Foulden: (a) *Strepsodus anculomnensis*, length 32 cm; (b) *Aetheretmon valentiacum*, length 7 cm; (c) *Acanthodes ovensi* length 7 cm; (d) *Carboveles ovensi*, length 14 cm; (e) *Rolfeia fouldenenis*, an early 'king-crab' (Xiphosura), length 7 cm; (f) *Belotelson traquairii*, a (probably herbivorous) shrimp, length 4 cm; (g) *Polylurida aenigmatica*, a 'worm' of some kind, length 5.5 cm.

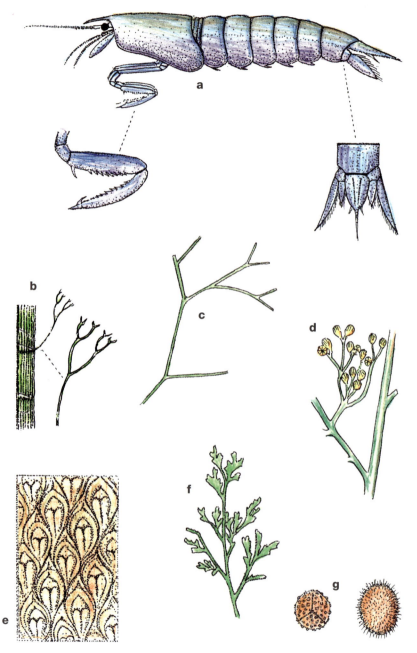

14.8 Crustaceans and plants from Foulden: (a) *Bairdops elegans*, a predatorial 'mantis shrimp', length 10 cm; (b) *Archaeocalamites*, a horsetail (sphenopsid), about natural size; (c) *Tristichia ovensi*, probably a seed-fern (pteridosperm), about half natural size; (d) *Ootheca* sp., a 'fern' with spore-cases (sporangia), about natural size; (e) *Lepidodendron aculeatum*, a club-moss (lycopsid) part of stem, about natural size; (f) *Telangium affine*, a fern (pteropsid), about natural size; (g) plant spores, much enlarged.

The Foulden Fish Bed was sampled bed by bed, and over an area of several square metres. In the course of excavation a large slab, some 2.4 by 0.6 m in surface area, and 0.3 m thick was lifted and brought back to the Museum's laboratory. Here it was stripped layer by layer, and all the recognisable constituents recorded so that changes in the faunas through time could be established. This gave an excellent opportunity to study the ecology of the lake, and the relationships of the animals to the environment and to each other.

The most diverse forms of life are the fish. Several of these were first described from Foulden, and some are not known from elsewhere. The commonest forms are the palaeoniscids, which are distinguished by having external head bones set in a characteristic pattern, diamond-shaped 'ganoid scales' on the body, and usually an asymmetrical tail. The four Foulden genera all have different kinds of teeth, modified for different diets. Acanthodians are usually rather small fish, distinguished by having the jaws not attached to the head, and peculiar fins, each of which consists of a stout spine set into the body, with a thin membrane behind. They are not as common as palaeoniscids, and nor are the small coelacanths which are also present at Foulden. There is no doubt, however, as to what were the top carnivores, the very large rhizodonts, which had wide gaping mouths and sharp teeth. These evidently fed upon other fish. They are commonest in the Shell Bed, deposited when the lake was at its deepest.

There are two species of crustaceans. One, the common *Belotelson*, is relatively small, looking like many shrimps living today, and was probably a scavenger or lower-level predator. The other, rare and much larger, was undoubtedly a predator. This genus, *Bairdops*, like the mantis shrimps (Stomatopoda) living today (and to which it may well be a precursor) has anterior appendages which could fold up like a jack-knife, impaling and crushing smaller prey with the many spines ranged along the inner edges.

There are many ostracodes, small bivalved crustaceans no more than a couple of millimetres long, and often occurring in swarms. There are also rare specimens of *Rolfeia*, a relative of the living 'horseshoe crab' *Limulus*, and known only from this locality, probably adapted for a benthic life on a soft substrate, but probably able to make short terrestrial excursions, or even climb up the lower parts of trees. It may have been washed into the lake. Otherwise the biota consist largely of typical Carboniferous plants, other than the wormlike *Polylurida*, known only from a solitary specimen.

The order of succession is very interesting. Above the basal *Myalina* bed, the large rhizodonts are present in the sediments, and following this again the two crustacean genera are present; *Bairdops* very likely preyed on *Belotelson*. The sediment in which these are found is somewhat bioturbated, in other words churned up by worms and other organisms living within. But at 30 cm above the base, there is a striking change; palaeoniscid fish appear and continue to be present until the lake silted up, but there are no more crustaceans and no further evidence of bioturbation. Towards the top of the Fish Bed is a layer consisting of very many small acanthodians, lying scattered on the bedding planes, and very probably this represents a mass-mortality horizon. No fewer than 173 specimens of *Acanthodes* were found to be present within the slab referred to, but not orientated in any way, so that there is no evidence of current sorting. So what might have been the controls acting to change these faunas through time? It is highly probable that the sediments were of brackish water origin, but as in many lakes or lagoons not far from the sea, salinity probably fluctuated. An extended period of rainfall, for example, would dramatically reduce the salinity of the water, and drought would increase it.

The following model seems to explain what happened during the life of the lake. At first water level was high, allowing very large rhizodonts to live in the lake. At this time the substrate was firm enough to support a fauna of bivalves, but it subsequently became softer. Infaunal creatures, however, could continue to live in the sediment at least for a time. The crustacean assemblage, ecologically balanced, probably was adapted to brackish water conditions, which excluded the fish. Following a rise in water level, palaeoniscids entered the lake, probably from a neighbouring freshwater pond, presumably adapted to a lower level of salinity, which the crustaceans could not tolerate. Following the mass-mortality event that killed so many *Acanthodes*, which may have resulted from a lethal rise in salinity or temperature, the lake returned to somewhat more saline condition, though one which both fish and crustaceans could tolerate, and in which they could all feed and flourish. These conditions prevailed until the lake finally silted up.

The Foulden lake has been well researched, though our knowledge is limited by only the one sampling point which is known at Foulden. We do not know how extensive it was, and the centre and margins may well have shifted during its existence. The depth, in absolute rather than relative terms at any one time, cannot be known. The Foulden site also yields typical Carboniferous plants, lycopods chiefly, but also some ferns and seed-ferns. They are mainly preserved as compressions, but also in one layer, the cellular tissues are still present in three dimensions. The Foulden sequence is only one of several successions containing fish–crustacean biotas in the Northumberland Cementstone Group, ranged along the southern margin of the Southern Uplands. Not all of these have been studied in detail. Yet the excavation of the Foulden site and its investigation by a team of experts has yielded a remarkably complete picture of this ancient lake and its surroundings, which lasted only a few hundred to a few thousand years, over 300 Ma.

CHAPTER 15

Volcanoes of the steamy wetlands

Significant changes to the geography of the Borders happened at about 359 Ma, coinciding with the start of the Carboniferous Period. These changes to the topography and river drainage patterns (Fig. 15.1) resulted from rapid crustal extension caused by plate movements taking place some hundreds of kilometres farther south. The extension involved some relative uplifts and subsidences. The latter created lowlands ('basins') in which river-borne sediments could accumulate. The Northumberland and Solway Basins (effectively one!) run WSW–ESE and essentially coincide with the collision zone between Avalonia and Laurentia, referred to as the Iapetus Suture. To the north of the Northumberland Basin, and separated from it by the Cheviot massif (Chapter 12) is the Tweed Basin. These basins contained semi-arid alluvial floodplains and, since 'the Borders' were now approaching equatorial latitudes (Fig. 15.2) the climate was becoming tropical.

The northern boundary of the Northumberland–Solway Basin is a flexure that can be followed intermittently from what is now the north side of the Solway near Kirkcudbright, east-north-eastwards for about 90 km as far as Cottonhope Head, some 12 km SW of the western limits of the Cheviot lavas. The flexure probably overlies a fault plane in the deeper crust that was sufficiently profound and deep-penetrating to have affected the underlying mantle. It is inferred that the resultant lowering of the pressure on already hot mantle rocks permitted them to start melting, with generation of basaltic magma and resultant volcanism. Activity of this type is referred to as 'intra-plate', far removed from any tectonic plate boundary. The nature of the eruptions and the compositions of the lavas contrasted strongly with the 'supra-subduction' magmatism responsible for much of the volcanic activity that attended Iapetus closure and that of the early Devonian described in Chapter 12.

As a result of being lighter (less dense) than the mantle peridotites within which they were formed, the basaltic magmas ascended. For most of the magmas, ascent is likely to have been checked some 30 km down when they reached the less dense,

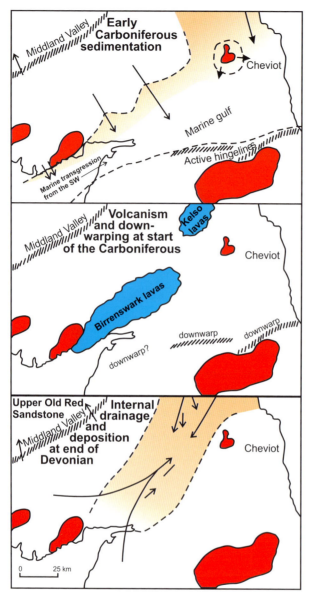

15.1 Map showing changes in drainage patterns from the late Devonian to the early Carboniferous (after M.R. Leeder).

feldspar-bearing rocks that form the crust. At this level the bulk of the magmas probably spread out laterally as sub-horizontal sheets. Magmas at this level would have cooled slowly, resulting in the onset of growth of olivine and pyroxene crystals. It is hypothesised that separation of these relatively dense minerals, perhaps by gravitational sedimentation to the floor of the sheets, gave the remaining molten magma renewed positive buoyancy, enabling it to continue ascent. The rising magma batches would also have continued to lose heat, with most of them cooling and solidifying as intrusions within the crust but some of them were successful in reaching the surface,

340 Ma
Early Carboniferous

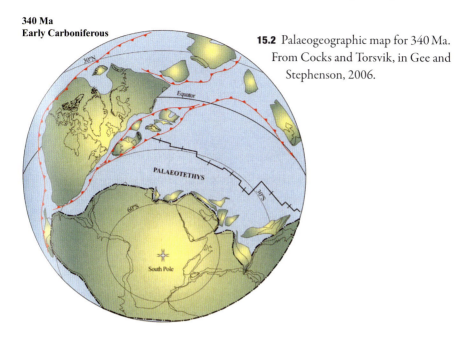

15.2 Palaeogeographic map for 340 Ma. From Cocks and Torsvik, in Gee and Stephenson, 2006.

erupting to produce a chain of volcanoes. At essentially the same time, closely comparable volcanism broke out away to the north-east in the Tweed Basin.

The faulting, down-sagging and volcanism associated with the Northumberland–Solway and Tweed Basins have features in common with the much more profound and famous rift valleys of East Africa. To a large extent eruption of basalt lavas was relatively passive, with an out-welling of fluid lava flows that contrasts with the more stodgy and explosive andesite and dacite lavas that characterised the Devonian volcanoes described in Chapter 12. It is thought that the lavas spilled out from their craters or fissures as broad, flat-lying tongues across the surrounding lowlands. Although the vigorous escape of dissolved gases in the magmas probably caused fountaining of incandescent lava droplets, the scarcity of strata composed of fall-out materials (tuffs) implies that explosive events were rare during these early basaltic phases. The youngest lava from any one volcano would have overridden, or flowed alongside, earlier flows from the same or neighbouring volcanoes so that an overlapping lava field grew, rather in the manner that molten candle-wax from adjacent candles might form a composite puddle.

The crust beneath the volcanic areas slowly subsided so that repeated activity at the numerous craters or fissures built up a lava sequence of intercalating and overlapping flows. The repose intervals between one eruption and the next may often have lasted hundreds or even thousands of years, during which silts and sands were commonly deposited across the lavas. In due course, these sediments were themselves buried beneath the products of the next eruption so that siltstone and sandstone layers are intercalated within the lava sequence.

The Birrenswark and Kelso lavas

The volcanic rocks to the west-south-west, along the edge of the Northumberland–Solway Basin(s), are known as the Birrenswark lavas whilst those in the Tweed Basin are referred to as the Kelso lavas. The two sets of lavas were contemporaneous but geographically separate. The Birrenswark lavas have an aggregate thickness of up to 90 m and appear to have spread over an area of approximately 1830 km² over the sandstones and lime-rich muds (marls) deposited by upper Old Red Sandstone rivers and lakes. The name derives from Birrenswark Hill, a prominent basaltic landmark east of the M74. The Birrenswark lavas form a low-lying ridge, crossed by the M74 and the railway some 3 km NW of Ecclefechan. Examination of the lavas is not easy because exposures tend to be small and widely separated. Until the early twentieth century, there would have been many small quarries in the lavas, but these have since been filled in or overgrown and exposures now are mainly provided by stream sections. The key sections are exposed in and around the Tarras Water valley, about 7 km NE of Langholm and 5 km NW of Newcastleton with some of the best exposure on Hartsgarth Hill.

The location of the individual volcanic vents is unknown, but they appear to have been sited along the ENE–WSW-trending lineament. However, by analogy with more modern situations, one can make a fair guess as to their probable size, shape and eruptive behaviour. Some of the lavas may have welled out of dilating fissures in the crust, i.e. from fissure volcanoes, whereas others, perhaps the majority, would have erupted from low-angled basaltic shield volcanoes rising to not more than a couple of hundred metres above the surrounding lowlands. In either case, the volcanoes are likely to have made relatively insignificant landscape features. It has been shown that the lava sequence thins northwards and westwards as well as to the south and east, so that the volcanic lineament was lens-shaped in cross-section and the eruptive fissures or vents almost certainly lay along the axis of maximum lava thickness.

The corresponding lavas to the north-east constitute the Kelso volcanic field (Fig. 15.1). Like the Birrenswark lavas, these flowed out across the plains, accumulating to a maximum thickness of 130 m. Post-eruptive subsidence again led to their subsequent burial beneath silts and sands of the early Carboniferous. Gentle tectonism towards the end of the Carboniferous and the start of the Permian, at around 300 Ma, down-folded both the lavas and the sediments. Kelso itself lies on the fold axis and as a result of later erosion, the lavas crop out in a V-shaped configuration to the south-west of Kelso, in which the 'V' opens to the north-east.

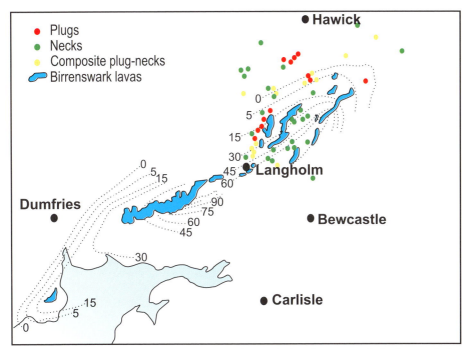

15.3 Map of the Birrenswark lavas. The lines ('isopachs') indicate overall thicknesses (after M.R. Leeder).

The thickness of individual lavas seldom exceeds a few metres. As they cooled, vesicles ascended through the molten lavas, concentrating towards the flow tops, whereas the lower, de-gassed, portions of the lavas were left vesicle-poor or vesicle-free.

Although basaltic magmas are not inherently explosive, each new batch of molten magma, at over 1000°C, inevitably encountered wet materials as they neared the surface. The heated (or indeed super-heated) waters ascended by virtue of their low density, and cooler (denser) waters flowed in to replace them, generating a convective flow much like that in a domestic central heating system. Such hot solutions circulated throughout the lava pile for tens or hundreds of years during and after their eruption and reacted with the earlier-formed minerals, degrading them to the greenish, platy silicate mineral chlorite and to carbonate and clay minerals.

Studies by Sergei Tomkeieff in the 1950s and '60s concluded that there were up to twelve separate flows between the Eden Water and the Tweed, but in some areas they can be represented by a single lava. Bedded tuffs occur between some lava flows, and in the Teviot section the timelapses between eruptions were long enough for some thick river-lain sandstone units to be intercalated. The Kelso lavas are best seen in the Blackadder Water near Greenlaw, in the Eden Water below Girrick, in the banks of the Tweed and Teviot rivers below Makerstoun House and Roxburgh respectively, and at Wooden Dean just south of Sweethope Hill.

Kershopefoot and Glencartholm volcanism

Continued slow subsidence of the Border region persisted throughout the early Carboniferous. Sedimentation was interrupted by two further outbreaks of volcanism, but by this stage, the semi-deserts of the late Devonian and earliest Carboniferous had given way to greener and better-watered landscapes.

The earlier of these outbreaks saw a revival of subaerial volcanism that yielded the Kershopefoot lavas along the northern margin of the Northumberland Basin. These occur as outcrops south and south-east of Langholm where they reach a maximum thickness of 30 m. The Kershope Burn locally defines the Scottish–English border and an old quarry at Kershope Bridge, together with a section alongside the Kershope Burn, afford the best places to see them. As for the Birrenswark and Kelso activity, the Kershopefoot lavas would principally have been erupted through small basaltic shields or cones.

Several million years later, around 340 Ma, came a final resumption of volcanism that produced the Glencartholm tuffs and lavas. These, up to 150 m thick, are best exposed in the River Esk about 5 km south of Langholm. Lava flows are scarce and the products are mainly tuffs comprising fragments of lava varying in composition from basalt to trachyte. (Trachyte is a variety of volcanic rock that is lighter in colour and density than basalt, considered in greater detail later in the chapter). The explosive nature of these eruptions, ensuring that the products were largely fragmental, undoubtedly resulted from the generation of steam from near-surface waters.

The sub-tropical landscape lay close to sea-level and was subject to marine incursions at times of high sea-levels, and rising magmas would commonly have encountered wet rocks, waterlogged sediments or standing bodies of water, and trapped pockets of water would have flashed into high-temperature steam, instigating dramatic boiler-type explosions. Such mainly steam-driven eruptions are referred to as 'phreatomagmatic' and give rise to distinctive volcano forms. These have shallow bowl-shaped craters enclosed by relatively low-rimmed walls and, left standing, these would have filled with water to form crater lakes. The term 'maars' is applied to such features, the name being taken from the Eifel district of western Germany in which such circular volcanic lakes of relatively recent origin are beautifully developed. The scenario diagrammatically shown is based on one of these (much younger) Eifel volcanoes, and is probably a fair representation of the sort of phreatomagmatic eruptions that typified many of these early Carboniferous volcanoes (Fig. 15.4). The sediments

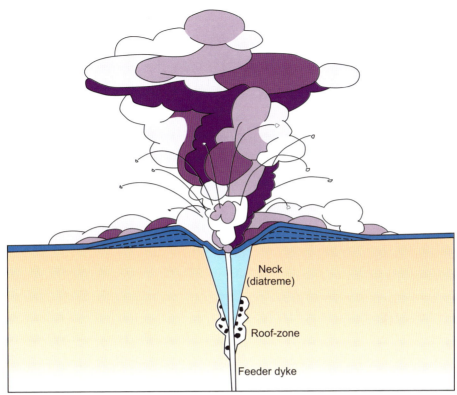

Neck
(diatreme)

Roof-zone

Feeder dyke

15.4 Schematic representation of a phreatomagmatic eruption with construction of a broad ('maar') crater surrounded by a low-altitude tuff-ring. A steep-sided neck ('diatreme'), filled with fragments of side-wall and juvenile volcanic rocks, develops below the crater. (After Suhr *et al.*, 2000).

associated with the volcanic rocks are not only remarkably rich in fish and invertebrate fossils, but they are also renowned for their plant fossils.

Volcanoes between Langholm and Duns

It is generally impossible to identify the precise volcanic centres from which any of the Birrenswark–Kelso, Kershopefoot or Glencartholm lavas and tuffs were erupted, but there is an abundance of intrusions and debris-filled vents between Langholm and Duns that indicate likely sites. They occupy an elliptical swathe (80 × 20 km), across the relatively uplifted region between the Birrenswark and Kelso lava fields. Most of the intrusions involved basaltic magmas, and having ascended from depths of over 60 km to levels not more than one or two kilometres deep, it is improbable that all should have halted and frozen before reaching the surface. Accordingly, the chances are that many of the intrusions did breach the surface to supply volcanic eruptions.

And in the case of the vents, we are certainly looking at sub-volcanic features. While the volcanoes must have presented very significant features, they would all have been dwarfed in comparison with the Cheviot volcano (Chapter 12).

Some of these volcanoes might have been responsible for some of the Birrenswark lavas, but several of the intrusions and vents cut through sedimentary strata that definitely post-date the Birrenswark (and Kelso) lavas. Accordingly these could have been contemporaneous with the Kershopefoot and/or the Glencartholm activity. All that one can be reasonably certain about is that these volcanoes represent a considerable spread in time (several million years) within the first ten million years of the Carboniferous Period.

Many of the volcanic stumps lie on the elevated watershed between the Hawick–Carlisle (A7) and Hawick–Newcastleton (B6399) roads. This area, mainly covered by moorlands and forests, sees few visitors other than shepherds and foresters and has been remarkably untouched by modern geological research. Much of what we know of this volcanic zone comes from the work of an indomitable woman, Rachel Workman McRobert, over ninety years ago (Fig. 15.5). Born Rachel Workman in Massachusetts in 1884, she studied geology at Royal Holloway College, London before taking her post-graduate degree at the Royal School of Mines. She published two papers of particular significance to Borders geology. The first, in 1914, concerned the intrusions and vents in the vicinity of Melrose (including the Eildon Hills, which will be considered later) while the second paper (1916) gave an account of the igneous rocks of Teviot and Liddesdale. As a result she, along with Gertrude Elles (Chapter 8), became in 1919 one of the first women to be elected as a Fellow of the Geological Society (Fig. 15.5). Marrying into the Scottish aristocracy she became Lady Rachel McRobert (sometimes spelling her name M'Robert).

In the mind's eye one can visualise an area of volcanoes dominating the landscape from Langholm towards Melrose and Kelso (Fig. 15.6). Probably they arose as vegetated hills that awoke from their slumbers after intervals of perhaps hundreds to thousands of years for an eruptive interlude that may have lasted only a few years at most. For any two volcanoes to have been simultaneously active would have been the exception rather than the rule. Looking at the profile of Rubers Law (just south of Denholm), for example, the uninitiated could be forgiven for thinking that it represents a recent, little-dissected volcano. This, however, is far from the case, and the conical shape derives from the differential erosion around a tough core (plug) of basaltic rock (Fig. 15.7).

Some of the intrusions are sills (sub-horizontal layers of igneous rock formed from magmas injected laterally between near-surface strata) but many have steep-sided margins and constitute more or less cylindrical masses to which the descriptive term

15.5 Rachel McRobert climbing in the Alps in 1906. Photograph courtesy of the McRoberts Trust.

Below **15.6** Diagrammatic impression of the volcanic zone across the Borders in the early Carboniferous.

'plug' is applied. These would have acted as near-vertical pipes through which magma arose to supply surface eruptions, and hence we can refer to them as sub-volcanic features. The plugs have diameters ranging up to 500 m, and formed when magma occupying them solidified to hard igneous rock.

The evidence left presents us with a not unfamiliar situation. Early lavas and tuffs are preserved (albeit sparingly and affected by secondary alteration), whereas the later extrusive products have been totally stripped by erosion, leaving only a view of

15.7 View of Rubers Law (424 m) beyond Denholm, from the north-west.

the (now blocked) pipe or conduit through which the magmas ascended. So we can find the extrusive rocks from the earliest activity, but are largely left guessing where they came from. At the same time we have somewhat later plugs and necks, clearly marking volcanic conduits, but with the lavas and tuffs that undoubtedly passed through them all gone (Fig. 15.8.). This partial record is frustratingly common in eroded volcanic terranes.

Many of the most prominent hills owe their origin to erosion-resistant igneous plugs that made them better able to withstand later grinding by the Pleistocene ice-sheets. Examples include Tinnis Hill, Hog Fell, Pike Fell, Bonchester Hill, Rubers Law and Maiden's Pap to the west and north-west of Newcastleton (Figs. 15.7; 15.9). Contrasting with the plugs are the vents or 'necks' choked with angular rock fragments that commenced as conduits blasted out by the high-velocity escape of gases. The Minto Hills provide good examples of necks (Fig. 15.10). In vertical section the necks are typically carrot-shaped, narrowing downwards as shown in Figure 15.4. Their infilling took place when gas venting ceased. Some of the fragments are

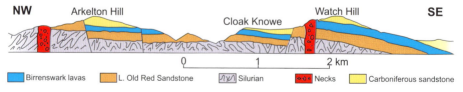

15.8 A NW–SE cross-section across Arkelton Hill to Watch Hill (after B.N. Peach and J. Horne). Upper Old Red Sandstone strata unconformably overlie highly-folded Silurian rocks and are overlain by lavas of the Birrenswark group, followed by lower Carboniferous sandstones. Two younger volcanic necks, occupied by fragmental rocks, are shown together with some minor faults.

15.9 Maiden's Pap (510 m) from the north-east (*c.* 12 km S of Hawick).

15.10 Minto Hills. Paired neck, *c.* 6 km NE of Hawick.

'juvenile', representing magmatic products of the volcano itself, whereas others came from the adjacent Palaeozoic sedimentary 'country-rocks'. Not infrequently one finds composite structures involving both a neck (usually the earlier component) and a younger plug. Examples are provided by Tinnis Hill, Rubers Law (Fig. 15.7), Blacklaw and Greatmore, in which an axial basaltic plug is surrounded by an earlier jacket of fragmental rocks. In such cases it is inferred that the slightly later ascent of basaltic magmas took advantage of the weakened passage provided by the neck. In these instances, an earlier phase of explosive eruption was succeeded by a less violent phase when rising the magma no longer encountered groundwaters.

To find a more nearly contemporary analogue for this volcanic region it is not necessary to go further than central France, where in the Massif Central a 50 km long linear belt of extinct volcanoes forms the Chaîne des Puys. These, with ages less than a million years, are much less worn down by erosion than the 350–340 m.y. old volcanoes of the Borders and they retain much more obvious volcanic morphologies.

Basaltic intrusions of the Kelso district

Tomkeieff reckoned that many of the outcrops in the Kelso area, formerly interpreted as lavas, are actually intrusions. Although the lavas and the bulk of the intrusions are virtually indistinguishable in composition, there are distinctive features by which the intrusions can be distinguished. Since the intrusions cut both Upper Old Red Sandstone strata *and* the early Carboniferous lavas, but not the overlying Carboniferous sedimentary rocks. Tomkeieff concluded that although they probably all postdate the surviving lavas, they could well have been related to younger volcanic activity, the extrusive products of which have been eroded away. His map (Fig. 15.11) shows over thirty intrusions, mainly basaltic but some trachytic.

Intruded magma tends to cool more slowly than that poured out on the surface, crystallising to rocks that are more coarse-grained than the lavas, with the component crystals commonly exceeding 1 mm across. Because of this textural difference, these coarser rocks are referred to as dolerite rather than basalt. So, the difference between dolerite and basalt concerns *only* the grain-size, *not* the chemical composition, and there is a complete gradation between the two! The doleritic intrusions tend to be more compact, with fewer cracks and fissures, tending to make them less prone to secondary alteration than the lavas. This renders them more attractive for use as aggregate, and the intrusions have been much quarried for roadstone. They are sometimes referred to as 'blue-stones' in contrast to the more altered greenish lavas. In brief, if an igneous rock in the area presents a craggy outcrop or if it has been quarried, it is probably an intrusive dolerite.

The intrusions tend to be either: a) plugs, crudely cylindrical and generally a few tens of metres in diameter, or b) sheets (sills) injected roughly horizontally between pre-existing sedimentary or volcanic strata. Plugs are abundant and are present at, for example, Hume, Hareheugh Craigs, Sweethope Hill, Queenscairn Hill, Fans, Mellerstain, Blinkbonny, Smailholm Tower (Fig. 15.12) and Bemersyde. As in the region between Hawick and Newcastleton, some plugs were intruded into fragmental (pyroclastic) necks as at Legerwood, Middlethird, Knock Hill and the Menteith Monument. Hareheugh Craigs is an unusually fine example of a composite plug. There are also many sill-like intrusions in the Kelso area, including those at Duns Law and Hexpath, as well as possibly Hallyburton Hill, Brotherston Hills, Ancrum

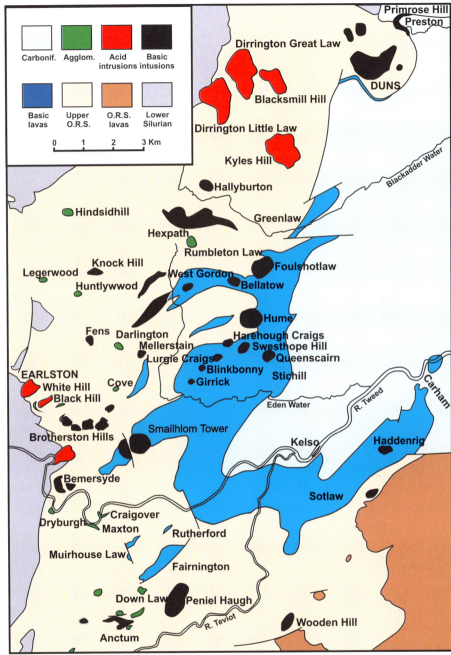

15.11 Map of Kelso lavas and intrusions, after Tomkeieff. The 'U-shaped' outcrop of Kelso lavas overlies upper Old Red Sandstone strata on its north, west and southern sides but is itself overlain by lower Carboniferous strata on its eastern side. A profusion of basaltic to doleritic intrusions (mainly plugs) cut through the lavas and the Old Red Sandstone. There are some (small) necks cutting the Old Red Sandstone as well as some trachytic intrusions (mainly sills) from Earlston in the SW to Blacksmill Hill and the Dirrington Laws in the NE.

15.12 Smailholm Tower, *c.*9 km west of Kelso as seen from the east, rising from knolls of basaltic lavas and intrusions.

and Wooden Hill. The Brotherston Hills probably represent a single, large sill which is being exploited for roadstone. The shape of some of the intrusions, however, is hard to tell; they could represent either plugs or sills. Such ambiguous cases include Peniel Heugh, Haddenrig and Lurden Law. Additionally there are numerous necks, including Rumbleton Law, Huntlywood, Cove, Dryburgh and Ancrum, filled with broken fragments (as agglomerates or tuffs). The Hexpath intrusion is noteworthy in that it contains fragments of mantle peridotite. This implies that the magma ascended particularly fast, ripping off and carrying bits of dense and refractory rock from mantle side-walls at depths between 70 and 30 km. A few plugs between Hawick and Newcastleton (e.g. Cooms Fell) also contain pieces of mantle peridotite. The xenoliths provide small fresh samples of the lithospheric mantle as it was in the late Palaeozoic. Although the age of these samples is unknown (apart from the fact that they are obviously older than the magma that transported them) they may well be representatives of the lithospheric mantle beneath Avalonia and possibly beneath the hypothesized Ordovician arc-complex.

Trachytes and rhyolites: what are they and how are they relevant to the Borders?

Trachyte has been mentioned with respect to the Glencartholm volcanics and the Kelso intrusions. Furthermore, trachytes and related rock-types form the best-known topographic feature in the Borders, the Eildon Hills (Fig. 15.13). Such rocks,

15.13 The Eildon Hills looking west from Scott's View (Bemersyde Hill). From right to left these are North Hill (404 m), Mid Hill (422 m) and Wester Hill (371 m). Red-coloured fields take their colouration from underlying upper Old Red Sandstone.

however, occur intermittently over 50 km of the broad swathe of volcanic rocks between Langholm and Duns.

However, before discussing in further detail their role in the Borders' story, a brief digression into igneous petrology is necessary. The majority of the lavas and intrusions discussed so far crystallised from basalt magma, the universally common product of mantle melting. When only a small proportion (less than around seven percent by weight) of the parent mantle rock melts, it produces magma in which there is a comparatively high ratio of alkalis (sodium and potassium) to silica. The early Carboniferous basalts are virtually all of this kind, the so-called alkali basalts. When substantial volumes of alkali basalt magma undergo slow cooling deep in the crust they experience the phenomenon of fractional crystallisation, in which the early crystals are poor in alkali elements. Accordingly the last dregs of magma left are corresponding rich in these components. These residual melts (and their solidified equivalents) are called trachytes. In brief, the composition of the melt changes during its crystallisation and the story does not necessarily end at trachyte.

According to the relative balance of silica and the alkali elements, the trachytes are either sufficiently silica-rich for the common silica mineral, quartz, to crystallise or sufficiently silica-poor for a silica-deficient mineral, such as nepheline, to crystallise. In the former case the magmas and their products are called quartz trachytes. But in latter case, the trachytes grade into another magmatic category, namely phonolites. In some cases, extreme fractional crystallisation of trachyte magmas produces even more silica-rich melts referred to as alkali rhyolites. Here we encounter another termino-logical hurdle, because the pale-coloured and typically very fine-grained products of these rhyolitic melts are usually called 'felsites' or 'felstones' in the older literature. However, for simplicity we will call the resultant rocks, rhyolites.

Trachytic, phonolitic and rhyolitic rocks appear intermittently within the early Carboniferous plugs, sills and vents. They have much in common with the trachytes and phonolites that occur in East Lothian and which are responsible for such well-known features as Traprain Law, North Berwick Law and Bass Rock, with which they are essentially contemporary. These residual rock-types can usefully be grouped under a single collective adjective, 'salic'.

Some of the rhyolitic rocks have peralkaline compositions. Peralkaline rhyolites have already figured in this geological saga, in Chapters 5 ('the Wrae volcanics') and 10 (some of the 'meta-bentonites'). When unaltered – as they tend to be in these relatively late Palaeozoic rocks – they contain the distinctive minerals aegir-ine and riebeckite. These have beautiful green and indigo-blue colours respect-ively when viewed in thin-section through a microscope. Although there are other occurrences of peralkaline salic rocks in the British Isles, for example in some much younger (Palaeogene) intrusions in the Hebrides, they are relatively uncommon.

If, on a conservative estimate, it takes eight to ten parts of basalt magma to produce one part of trachyte, the comparative abundance of trachytes (plus associated phon-olites and rhyolites) north-east of Langholm implies that remarkably large volumes of basaltic magma were involved in their generation. The requisite differentiation processes probably occurred in magma chambers deep in the crust. In nature there is a complete spectrum of compositions between basalt and trachyte (and thence towards either rhyolite or phonolite).

Salic and basaltic magmas contrast markedly with respect to their behaviour which, in turn, leaves its mark on the Borders landscape. Salic magmas have higher contents of potentially volatile components dissolved in them than do their parent basalt magmas. Of the many volatile constituents, water and carbon dioxide are by far the most abundant. And, as explained in earlier chapters, when the magmas rise towards the surface, the pressure on them from the mass of overlying rocks diminishes and their capacity to retain their volatiles falls likewise, and the volcanic gases come out of solution and escape into the atmosphere. Among the consequences are that the loss of these gases promotes crystallisation, causing the magmas to become increas-ingly sticky and viscous. This means that there is less chance that they will erupt on the surface but a greater chance that they will crystallise as shallow crustal intrusions. A fair number of these are to be found in the volcanic zone north of Langholm. In a number of cases these are sub-horizontal tabular intrusions (sills) squeezed out along

15.14 Skelfhill Pen (532 m) from the east. The high-ground (*c.*12 km SSE of Hawick) is formed by a complex suite of trachytic intrusions, ranging from silica-undersaturated to silica-oversaturated.

planes of weakness in the sedimentary strata. Like the basaltic intrusions, the salic rocks tend to be more resistant to erosion than the enclosing sedimentary rocks so that they are left standing out as hill-forming features. Apart from sills, salic rocks occur as plugs and dykes and there are also some necks that were associated with trachytic rather than basaltic magmas. Just as there are basalt-related necks cut by basalt plugs, there are also examples of trachyte plugs within necks (e.g. Wether Law, another below Garlin Tooth and on the ridge opposite Penchrise). Furthermore, there are large necks in which the pyroclastic filling (agglomerate and tuff) is cut by both basaltic and trachytic intrusions as at Tudhope Hill and Greatmore.

Skelfhill Pen, about 12 km SSW of Hawick, is remarkable not only for the relative abundance of trachytic rocks but for their compositional range; they vary from silica-oversaturated (quartz-bearing) to silica-undersaturated (nepheline-bearing) types over a distance of a few hundred metres (Fig. 15.14). While it awaits a detailed study, the probability is that the quartz-bearing trachyte magmas acquired their higher silica through reacting with, and assimilating, relatively silica-rich crustal rocks, whereas the silica-undersaturated magmas ascended with little or no crustal interaction. A broad phonolite dyke forms the summit of Skelfhill Pen, while north of the summit is what McRobert reckoned to have been the main volcanic vent. A set of NE–SW-trending trachytic dykes connects Skelfhill Pen with another trachytic occurrence 9 km further to the south-west, at Pikethaw Hill just west of the A7.

The Eildon Hills

The Eildons must rank as the most famous topographic feature of the Borders. The three Eildon Hills (North, Mid and Wester), varying from 371 to 422 m high, inspired the Romans to name their nearby camp at Newstead, Trimontium. The hills are dominantly of trachtyic and rhyolitic rocks. Recent radiometric dating on the Eildons gives us an unusually precise age of 352 ± 1.43 Ma, putting them very early in the Carboniferous and roughly contemporaneous with the Kelso lavas.

The Eildons present a terraced profile that McRobert interpreted as due to the weathering out of several intrusive sheets of trachyte and 'felsite'. The intrusions appear as sill-like injections along the unconformity plane separating the steeply folded Silurian 'basement' from the near-horizontal Upper Old Red Sandstone strata. An extensive sill of quartz trachyte (possibly the earliest) underlies all three hills. An isolated 'outlier' of the same rock-type at Bowdenmoor, 400 m to the west of the Eildons, shows that the original extent of the sill was considerably greater. Further confirmation comes from other outlying slivers north of North Hill and south-east of Chiefswood Hill (about 1 km NW of the Eildons). The basal sill on North Hill is overlain by another sheet of trachyte that differs in texture from the underlying one. However, the trachyte forming the highest unit on North Hill could either be part of the basal sheet or a quite separate intrusion.

Wester Hill and much of the lower part of Mid Hill are made of riebeckite-bearing felsite (i.e. peralkaline rhyolite) and, judging from the topography, possibly consist of two parallel sub-horizontal sheets (Fig. 15.15). An outcrop on the south-western slopes of Wester Hill exhibits well-developed columnar jointing. Mid Hill is clearly the most complex part of the Eildons, consisting of up to five separate trachytic units. Towards its summit, on the western side, is a trachyte sheet with a somewhat more 'primitive' composition (implying hotter magma with slightly more magnesium and calcium than the others). Capping the hill and forming a veneer down its eastern side is a riebeckite-bearing trachyte, similar to those at Pikethaw Hill and Skelfhill Pen. The Eildon rocks have been affected by warm, hydrous solutions that permeated them, probably during their cooling history. The consequent oxidation resulted in the rocks acquiring pinky-brown tints from haematite staining.

A small neck forms Little Hill on the western flanks of the Eildons, which McRobert described as being composed partly of basaltic tuff and partly of basaltic intrusion(s). The green, grassy slopes below Little Hill contrast with the more barren or heather-clad flanks that characterise most of the Eildons. This undoubtedly results from the different composition, the basaltic rocks weathering to yield soils that are richer in magnesium and phosphorus (among other components) and changing the vegetation.

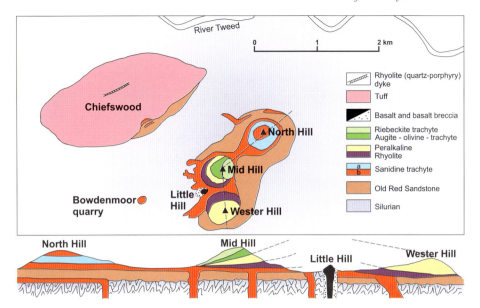

15.15 Map of the Eildon Hills and the Chiefswood Neck. Also a schematic cross-section (approximately NE–SW) through the Eildons showing McRobert's interpretation of the structure. (After McRobert, 1914.)

It would appear that the Eildon sills were formed from relatively silica-rich magmas, produced by a differentiating basaltic magma chamber at depth. They then spread out laterally as shallow intrusions at near-surface depths before being followed by the ascent of basaltic magma that may have penetrated through, to erupt at the surface. If this was so, Little Hill may be construed as marking the site of an eroded volcano.

Other trachytic and rhyolitic occurrences in the Melrose–Greenlaw–Duns area

Apart from those composing the Eildon Hills and its surroundings, there are several other salic intrusions in the Melrose area. These may be roughly contemporaneous with the Eildon intrusions and all of these magmas could have arisen from the same deep crustal source. These trachyte and quartz trachyte occurrences include White Hill, Black Hill, Bemerside and Whitlaw Hill. A number of salic dykes cross-cut the area, with trends typically parallel to the Caledonian 'grain' of the underlying Silurian.

A grassy hill between Dingleton and Darnick, west of the Eildons, is known as Chiefswood. It is the site of an unusually large (*c*.800 m x 350 m) vent, elongate ENE–WSW, filled with (almost unstratified) tuff. The latter comprises fragments of the Silurian and Devonian wall-rocks and scarcer igneous rocks, mainly trachyte. The trachytes suggest that the vent postdated some (or all?) of the Eildon intrusions. The vent is bisected by an ENE–WSW-trending rhyolitic dyke which, like the elongation of the vent, reflects the general Caledonian grain of the Silurian country-rocks.

The Chiefswood vent and some other small necks in the vicinity are thought to have resulted from rhyolite magma encountering water at or near the surface. The resulting steam explosions would have produced a volcano with a broad low profile much as illustrated in Figure 15.4. The igneous fragments also include basalt and rhyolite. Analysis of the latter shows it to have come from a peralkaline magma and, since the cross-cutting dyke appears to have a similar composition, it is a reasonable presumption that the same magma type persisted at shallow levels from the pre-eruption to the post-eruption stages.

The fresh tuff from the Chiefswood vent has a mottled brownish coloration and makes an attractive (and easily worked) building-stone (Fig. 15.16) and there are (or were) several quarries cut into it (Fig. 15.17). The tuff was used extensively in the building of Melrose Abbey as well as other old buildings and walls in the town.

The most northerly expression of the early Carboniferous magmatic activity in the Borders is a group of intrusions cutting the lower Old Red Sandstone conglomerates between 5 and 10 km west of Duns. These form Dirrington Great Law, Dirrington Little Law, Blacksmill Hill and Kyleshill, rising to heights of up to 400 m. The intrusions which, like those of the Eildons, are sills (or their allied form, 'laccoliths') last studied in the late 1920s. The first three hills are made of what were then called

15.16 Cut and polished surfaces of a sample of the tuff from Chiefswood Vent. The very black and dark grey fragments are from the Silurian 'country-rocks'. Small brick-red fragments are of iron-stained silica (jasper) from vein deposits. The majority of rock fragments however are of trachytes – similar to those forming the bulk of the nearby Eildon Hills whilst the very pale, cream-coloured fragments are of peralkaline rhyolite.

15.17 Jointed tuff in a disused quarry on the NW side of Chiefswood Hill, between Darnick and Dingleton.

'felstone' but which would now be called peralkaline rhyolite, and probably all constituted parts of a single intrusion. Compositionally they are comparable to those of Wester Hill in the Eildons. Apart from relatively large (up to 4 mm) early-formed crystals of feldspar and quartz, the rocks have a fine-grained matrix which possesses a texture in which microscopic feldspar and quartz crystals sit radially, forming tiny spheres. Such a texture is typical of that formed when a glass (which is a super-cooled liquid) crystallises. All glasses are inherently unstable and are ephemeral in terms of geological time. The inference is that the magma in question, carrying a small proportion of quartz and feldspar phenocrysts, was injected into the conglomerates, where it rapidly lost both heat and dissolved water and carbon dioxide, causing it to congeal as a natural glass. The glass (obsidian) would subsequently have crystallised ('devitrified') to form the observed textural features.

The Kyleshill intrusion is somewhat different and is a quartz trachyte rather than a rhyolite. Despite this difference, the suggestion that all four salic masses were components of, if not a single intrusion, then a closely related set of sills, appears highly probable.

The presence of all these early Carboniferous salic intrusions and related necks suggests that an unusually large body of basalt magma must have underlain the area, to have generated such a relatively large volume of residual magmas.

Summary

Early Carboniferous magmatism affected a zone stretching from near Kirkcudbright to the vicinity of Duns. For the sector between Kirkcudbright and Langholm, the zone essentially follows the line of the 'Iapetus Suture'. But beyond the Langholm area, the zone swung into a more north-easterly trend, transgressive to the underlying Lower Palaeozoic 'accretionary wedges' (Chapters 5 and 6). Along this broad volcanic zone magmatism was longer-lived and notably bimodal in character, involving both basaltic and salic magmas, with little in the way of intermediate compositions.

Whereas some of the many plugs between Langholm and Selkirk might represent eroded feeder pipes for the Birrenswark lavas, other plugs (and necks) cut through early Carboniferous sedimentary strata and are thus demonstrably too young to be related to the Birrenswark volcanism. On a similar argument, many of the intrusions cutting the Kelso lavas cannot represent the conduits through which the lavas were supplied. We may reasonably presume, however, that the basaltic intrusions underlay volcanoes whose extrusive products have been eliminated by later erosion.

In some cases, basaltic plugs were intruded through pyroclastic necks that are circular or ovoid in plan. Remembering that we are considering a sub-tropical lowland terrain in which rivers, lakes and lagoons abounded, the plug and neck associations suggest that magma batches commonly encountered water at near-surface levels, giving rise to steam-driven phreatomagmatic explosive eruptions. Further ascent and intrusion of magma probably caused elevation of the ground and recession of the waters responsible for the steam generation. More quiescent eruption of the basaltic magma could then take place, probably ponding as lava lakes as well as overflowing the crater rims, causing lavas to flow down the volcano flanks. Repeated extrusion of rather fluid basalt lavas could then have built up volcanoes with low-angled flanks ('shield volcanoes') rising to modest heights.

Basaltic eruptions contrast sharply with trachytic and other salic ones. The more volatile-rich salic magmas tend to freeze up near the surface as shallow intrusions (e.g. those forming the Eildon Hills) but, where they do burst through to erupt, they typically produce explosive eruptions, yielding fast-flowing and potentially lethal 'ash-flows'. Such flows are composed of disrupted shards of magma and crystals grown in it, plus bits and pieces of already solidified rock ripped off during eruption, buoyed up in a matrix of very hot gas. The ensemble has relatively high density and, by virtue of its lubricating gaseous matrix, very high fluidity. It can move downslope under gravity very much as does a snow avalanche. As the ash-flow loses heat, momentum and gas content it collapses to produce a rock-type known as ignimbrite.

In contrast to this behaviour, residual trachyte magmas that have lost the bulk of their gas contents can be sluggishly extruded as viscous lava. Typically these are thick, slow-moving flows that can travel no great distance and behave rather in the manner of toothpaste. While we now see no traces of either ignimbrites or degassed lava masses ('domes'), we may expect that they were closely associated with the principal salic centres such as Skelfhill. The Chaîne des Puys in France provides us with some excellent examples, as in the Puys de Dôme and the Dôme de Sarcoui (Fig. 15.18) of the sort of volcanic morphology that would have risen above the vegetated wetlands of early Carboniferous Borders country.

Some later magmatic arrivals in the Borders

Although after the early Carboniferous eruptions, volcanoes never again added excitement to life in the Borders, there were two subsequent magmatic events that left their indelible mark. One of these occurred some 50 m.y. later in the closing days of the Period, whilst the other occurred relatively recently (a mere 58 Ma) in the Palaeogene.

In the earlier event, at around 300 Ma, large volumes of basaltic magma were generated in the upper mantle, possibly beneath Scandinavia. Although the reason for this melting remains controversial, one current hypothesis is that it was due to a great mass of convecting hot rocks rising through the mantle and partially melting as it neared the surface. While the question 'why' cannot be answered, there is no question about the result. The basalt magma invaded northern England and southern and central Scotland as east–west trending dykes, many of which had widths measurable in tens of metres. It is inferred that magma, supplied through these dykes, fed into two huge sills or sill complexes, one in northern England (the Whin Sill) and the other forming the Midland Valley Sill complex in the Forth valley region. Although there are some insignificant representatives of this great dyke swarm in the Borders, they form no landscape features.

The heat from the Whin Sill was responsible for a major convective disturbance of the waters contained within the shallow crust. In other words it gave rise to extensive circulation of hydrothermal solutions, and some alteration of the rocks in the Cheviot massif has been ascribed to these warm percolating waters.

After another 250 m.y. had elapsed a further mantle upheaval in the northern hemisphere led to the stretching and rupturing of the vast continent of Pangaea (which included much of Asia, Europe and North America). This saw the commencement

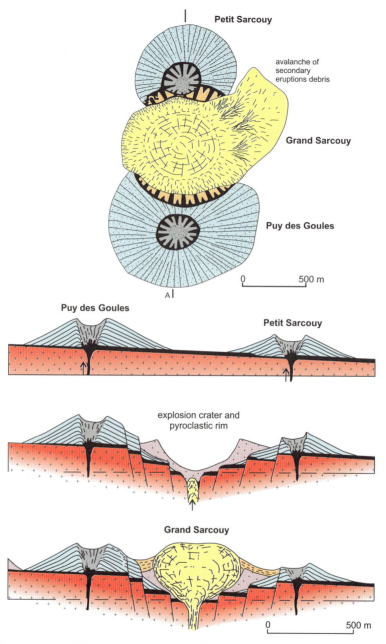

15.18 A map of Grand Sarcouy and adjacent volcanoes, Chaîne de Puys, Auvergne, Central France. Cross-sections showing three evolutionary stages in its formation are shown below. A phreatomagmatic explosion crater was generated between two adjacent basaltic volcaoes, Petit Sarcouy and Puy des Goules. Viscous trachyte magma was then extruded through its conduit to form a large bulbous dome (Grand Sarcouy). Volcanic land-forms of this type are inferred to have developed in the Borders in the early part of the Carboniferous Period. (After de Goer *et al.*, 1991.)

of the infant North Atlantic Ocean as the broken continental portions pulled apart. This situation was thus a replay of the events that saw the super-continent of Rodinia break up to give birth to the Iapetus Ocean.

The preliminary stretching that heralded the North Atlantic began around 60 Ma while the actual plate separation took place some 5 m.y. later. The stretching, together with the addition of mantle heat, caused extensive magmatism throughout the Hebridean and Northern Irish sectors, resulting in the growth of large volcanoes, for example on what are now Skye and Mull. The magmatism also saw generation of swarms of basaltic dykes, with a predominant NW–SE trend. Whilst these were mainly bunched around the big volcanic centres, others were propagated far to the south-east. Some of the most remarkable examples were associated with the Mull volcano, extending across the Highlands, central and southern Scotland and even into northern England. Those that concern our story cut right across the Borders, almost at right angles to the ancient Caledonian grain inherited from the Iapetus closure (Fig. 15.19). The two largest are the Acklington and Eskdalemuir dykes, each with widths of 10–30 m, but locally attaining a width of 50 m. The Acklington dyke can be traced intermittently in the field, or from aeromagnetic mapping, for over 90 km

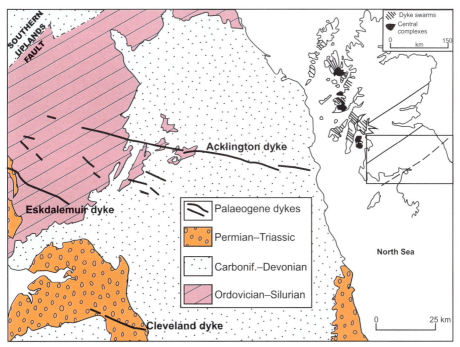

15.19 Simplified geological map of the Borders showing the Palaeogene dykes cross-cutting the Lower and Upper Palaeozoic rocks. The inset map shows the relationship of these dykes to the eroded Mull volcano and its dyke swarm.

from close to Hawick, past Bonchester and across the southern crops of the Cheviot lavas into Northumberland, where it is last seen towards the coast near Amble.

The Eskdalemuir dyke can be identified for about 55 km, cross-cutting the Ordovician and Silurian of the Eskdalemuir Forest, across the line of the M74 north of Beattock and on towards Langholm. Several other members of the Mull swarm cross the Borders but are only traceable for short distances. These are found north-east of the Eskdalemuir dyke, in the vicinity of Abington, Crawford and Elvanfoot.

CHAPTER 16

The iceman cometh

Southern Scotland during the Mesozoic and Tertiary

The Southern Uplands form a high dissected plateau whose relief is largely of ancient origin. The valley of Lauderdale, for example, as we have seen, was carved out at some time between the late Silurian and early Devonian, and still retains part of its Old Red Sandstone filling. We know very little of what happened in our region from about 240 Ma until comparatively recent times. In particular, we have no clear perspective on how much rock was eroded from the Borders from the early Carboniferous onwards, since there are no post-Carboniferous sedimentary rocks in the Borders.

Yet the immense global changes outside our own area must surely have affected the Borders in some important ways. In the Cretaceous, for example, there were no polar ice caps and there is great evidence of global flooding and raised sea-levels of more than 100 m. During the time when the Upper Cretaceous chalk was deposited, exposed land surfaces were much less extensive than they are today. We have no Cenozoic sedimentary rocks in the Borders, and what happened during that time remains uncertain.

The discovery that the British Isles had been glaciated

The earliest suggestion of glacial activity in Scotland was made by Robert Jameson, who as long ago as 1827 noted that boulders, often encountered on rock surfaces or moorland, are commonly of a quite different rock-type to that on which they rest. According to Jameson, these 'erratics' could well have been transported by glaciers which had then melted. But it was not until 1840 that real evidence for the former glaciation of Scotland was forthcoming when the renowned Swiss scientist, Louis Agassiz, visited Edinburgh. With his great experience of Alpine glaciers, he was able to pronounce, when inspecting horizontal striations on the south face of Blackford

Hill, 'This is the work of ice'. The idea became popular and in 1863 Archibald Geikie published a definitive 190-page paper entitled *On the phenomena of the glacial drift in Scotland*, as a result of which the concept of former glaciers in Scotland became remarkably topical. There remained doubters, but by the end of the nineteenth century the glacial theory became fully accepted, especially following the third edition (1894) of *The Great Ice Age* by James Geikie, who was Archibald's younger brother.

An agreed mechanism for the Ice Age, however, was long in coming, and it was not really until the 1960s that the reasons for the growth and decay of the ice sheets were properly understood. In late Cretaceous times there were no ice sheets, and the early Cenozoic was likewise very warm. But then around the Eocene/Oligocene boundary, some 35–40 Ma, a long and sustained period of global cooling set in. This was by no means uniform; temperature levels fluctuated considerably, but there was a net trend to increasing coldness intensifying toward the present time. The reason for this was that the Antarctic continent drifted from an originally temperate position to reside over the South Pole. Snow forming on this land surface built up into thick ice, and the reflection of the sun's rays on this new white wilderness made it colder still. The rhythmic growth and decay of the ice sheets was directly the result of Milankovitch cyclicity, as discussed in Chapter 7. We are at present living in an interglacial period; such periods are relatively short by comparison with glaciations, and there can be no doubt that the ice sheets will return. At least seventeen glacial episodes have been recorded in marine sediments around Scotland, and we can surely expect more. Of these only the latest, the Devensian, as it is known, left a good record in the Scottish Borders, although traces of the preceding (Ipswichian) interglacial, and the earlier (Anglian) glacial period are found in other parts of the British Isles.

The first clear evidence of global cooling comes from Antarctica, which because of its polar position is especially sensitive to climatic change. There seems to have been a temporary spread of the ice in the late Oligocene, some 25 Ma, followed by a warming event. But there is no doubt that permanent ice sheets were forming in Antarctica some 14 Ma and by 2.6 Ma icebergs were dropping rock debris in the North Atlantic. By 2 Ma Arctic microfossils displaced temperate forms in the North Sea and by 0.8 Ma real ice sheets were forming over upland Scotland, pushing eastwards into the North Sea. The Ice Age had truly arrived.

The base of the Quaternary Period is now taken as 2.6 Ma, though its effects in the Borders are not greatly clear before about 0.8 Ma. In order to analyse the events in real detail, scientists examine cores taken from late Cenozoic and Quaternary sediments from the deep oceans. These are rich in microfossils, especially planktonic foraminiferids, and not only do these assemblages themselves oscillate regularly through time, they also preserve a record of stable oxygen isotopes which can be used in precisely measuring the temperatures that existed when the foraminiferids were alive. The theory is quite simple. Atmospheric oxygen, and that dissolved in the sea, exists in two isotopes, the lighter O^{16}, and the heavier O^{18} which has two more protons in its nucleus. In warmer conditions the lighter O^{16} evaporates more readily, so that the oceanic waters have a somewhat higher proportion of O^{18}. In colder periods there is less evaporation and the ratio of O^{16} to O^{18} is somewhat lower. The isotope ratios present at the time a foraminiferid was living are preserved in the shell, and by working horizon by horizon through a core, the stable isotope levels can be assessed using a mass spectrometer. The cores show pronounced and regular oscillations resulting from Milankovitch cyclicity, just as we saw with the radiolarian cherts discussed in Chapter 7. So we have a regular alternation of warmer and colder periods, the latter becoming more intense with time. Also, for some reason, the record picked up the 40,000 year cycle before about 0.8 Ma, but after that the 100,000 year cycle becomes prominent.

Evidence of glaciation in the Borders

It was only the latest glacial period, the Devensian, that has left substantial traces of ice movement, evidence of earlier glaciations having been erased by the Devensian ice sheets. In the Lammermuir, Moorfoot and Lowther hills, for example, the land surfaces have been scraped clean by Devensian ice and the hills are smooth and rounded. The evidence of former glaciations comes from several sources. which include the sculpting of the land surface by the ice into characteristic forms, the kinds of sediment left behind as the ice melted and retreated, collectively known as glacial drift and glacial erratics, the large boulders dropped by the melting ice.

Glacial landforms

The most typical large-scale features resulting from ice-flow are U-shaped valleys, carved out by slowly moving glaciers, morphologically quite different from the V-shaped valleys cut by rivers. All the U-shaped valleys we have in Scotland were originally V-shaped river valleys, but the immense erosive power of the slowly moving ice has moulded them to their present form. Typically, U-shaped valleys have 'hanging valleys' up on their sides, where the original courses of the pre-glacial small tributaries have been truncated and the slope below them is much steeper than it originally was. The best-known examples are from the north and west of Scotland, but in the Borders the Manor Hills provide some fine examples. Of these the most spectacular is the valley in which the Talla Reservoir lies (Fig. 16.1). When seen from the steep eastern end (itself a kind of corrie marking the limit of the valley glacier) it presents the typical U-shaped form, complete with hanging valleys and cascading waterfalls.

16.1 Talla Reservoir, looking westwards. A typical U-shaped valley, sculpted by glaciers.

Likewise the Megget valley is U-shaped; though, occupied as it is by a deep reservoir, the form is less immediately evident.

A typical glacial landform, common in south-east Scotland, is known as 'crag and tail', this term having been introduced by Sir James Hall in 1815. Such structures as the Castle Rock and Blackford Hill in Edinburgh, and North Berwick Law and Traprain Law in East Lothian, are characteristic, as is the crag upon which Smailholm Tower (Fig. 15.12) is located. They are all Carboniferous-age igneous intrusions in which the western faces are very steep and craggy, while the eastern slope declines quite uniformly into an extended tail. These land-forms are the direct result of ice moulding, but they tell us also that the ice must have been moving in an easterly direction, as confirmed by many other sources.

Glacial drainage channels

The end of the last glaciation was sudden and dramatic. Within a single human lifetime the temperature rose many degrees, and over a kilometre thickness of ice melted within a relatively few years. The vast quantity of meltwater carved out channels in the land surface as it flowed, generally modifying older features. The long, straight valley of the Windy Gowl, near Carlops, is just such a channel, probably originally cut sub-glacially, and then modified by surging meltwater along the weak and easily eroded line of the Pentland Fault. Such meltwater channels are often quite small-scale erosional features, usually cut into drift, and often no more than a few metres deep,

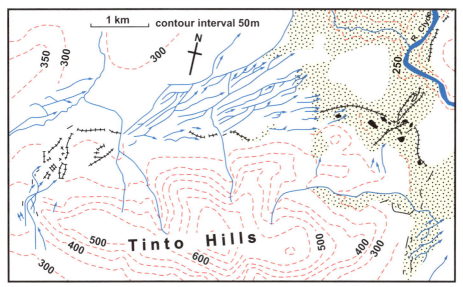

16.2 Glacial meltwater channels and other features north of the Tinto Hills. Eskers shown as dark lines with arrows, kettleholes in black, fluvioglacial deposits shown as dots (after Sissons).

that resulted from rushing torrents of water coming from a decaying ice-front. In Scotland there are literally thousands of these, and within the Borders some of the best examples are to be found to the north of the Tinto Hills (Fig. 16.2). Here there is a magnificent system of meltwater channels and the winding ridges known as eskers (defined below), which form part of an extensive drainage system extending from the Clyde to the Lammermuir Hills. The generally north-easterly direction of these is consistent with the general direction of Southern Upland ice flow.

In East Lothian the meltwater channels are much larger, some being more than 20 m deep. These drained to the east, following the Southern Upland Fault line for some tens of kilometres and then, following the eastern contours of the hills, they swing round in a south-easterly direction. The best place to see them is in a 4 km stretch, some 8 km SW of Dunbar. Here there is a long, sinuous meltwater channel, cut in terraces of drift, locally over 200 m broad, that extends eastwards from Deuchrie Farm (NT 623714). As well as this main channel there are eskers, kames, and kettleholes (again, see below); altogether it is a fine example of the complexity that such systems can achieve (Fig. 16.3). In the same general area, at Kidlaw Farm (NT 509642) is one of the largest glacial erratics in all of Scotland, a hill of broken limestone 500 by 400 m in extent, which has actually been quarried.

When glacial ice melts, the high mountain tops become clear first while the valleys commonly remain blocked by vast ice masses, allowing meltwaters to accumulate

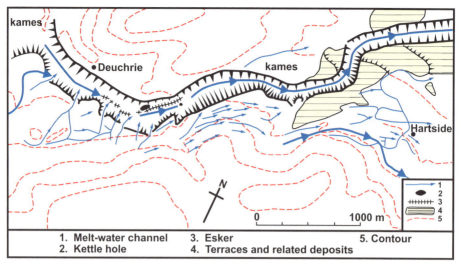

16.3 Glacial meltwater channels, eskers, kettleholes and terraces between Deuchrie and Hartside Farms, East Lothian (after Sissons)

16.4 Overflow channel near Aikengall, East Lammermuir Deans.

behind and/or on top of the ice plug. With further melting the dammed water eventually breaks through, carving out substantial 'overflow channels' as it does so. The example shown in Figure 16.4, cut through Lower Devonian conglomerates in the East Lammermuir Deans, is one of these.

Glacial deposits

When ice sheets recede they leave behind characteristic kinds of sediment on the surface (Fig. 16.5). The outermost extent of the former ice sheet is typically marked by a terminal moraine. This is an extended, often irregular or lobate line of boulders and debris, which fell from the front of the ice sheet when stationary. Where the ice sheet retreated in halting stages, with pauses in between, a series of recessional moraines resulted. Behind these, the formerly glaciated ground is usually covered with till or boulder clay, which is simply a clayey sediment, replete with boulders, left behind by the ice. In some cases this forms elongated mounds known as drumlins, in which the blunt end is orientated upstream. Drumlins occur across a range of scales and there are even examples tens of kilometres long. They are known to have formed under thick ice, floored by till, and they give clear evidence of glacial streamlining. There are some very fine drumlins in the Cheviot Hills and in the Tweed Valley in Berwickshire (see below). Finer sediment, in the form of fluvioglacial sands deposited by meltwaters, may form deltas or outwash plains.

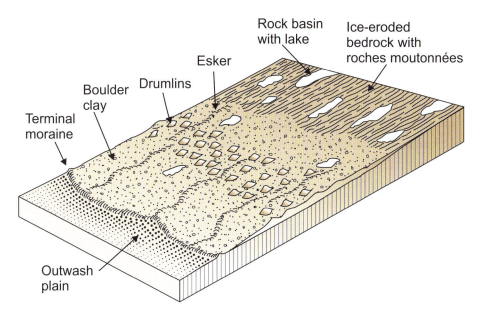

16.5 Typical features formed on low ground during ice retreat (after Holmes)

Eskers are long ridges of material left behind by rivers which flowed under the ice (Fig. 16.2), and they are often associated with meltwater channels. The complex esker and channel system, already referred to north of the Tinto Hills, provides an excellent example, and there are also fine ones in East Lothian at Eddleston, north of Peebles, and Bedshiels, north of Duns.

Kames are isolated or clustered mounds of boulders, which were transported by temporary rivers flowing on the surface of the ice, and which tumbled down at the ice-front to form alluvial cones. Some kames also contain interbedded sands and gravels, and such deposits often show small-scale faulting. Kettleholes are irregular hollows formed by stagnant masses of ice which took a long time to melt. Both kames and eskers, along with meltwater channels, form an almost continuous belt from the Clyde to the east coast, and were derived from melting ice which formerly covered the Southern Uplands. Exposed rock surfaces were commonly sculpted by moving ice into *roches moutonnées*, where the main bulk forms a smooth polished mound, but the rock face pointing in the direction of ice movement is sharp and irregular, where blocks have been plucked from the surface and taken away by the ice. They have formed in the opposite direction to a crag and tail. Unlike the other features described here, *roches moutonnées* only form during active growth and ice movement. These are not common in the Borders, but some good examples are to be found at the head of the Dalveen Pass.

Subglacial ice-stream landforms

In the drainage basin of the River Tweed, between the Cheviot and Lammermuir Hills, there is a swathe of exceptionally elongated subparallel mounds, basically drumlins, but much longer and narrower than usual. Some are large enough to be classified as mega-drumlins In the western part of the area the swathe is broader, some 40 km across, and these features are aligned in a north-easterly direction. Further east the zone with these unusual landforms narrows to some 20 km and wraps round the northern edge of the Cheviots to turn south-eastwards. The tract as a whole is at least 60 km long, and may have extended further into what is now the North Sea. These landforms are clearly seen along the A697 (from Carfraemill to Wooler) between Greenlaw and Coldstream, where the road cuts through them, and where their scale can be appreciated. These elongated mounds result from the passage of ice-streams, which are bodies of fast-flowing ice that move at a much greater velocity than the surrounding ice. The ice-stream rode rapidly over a basement of deformable till, sculpting and streamlining it into the shapes we now see, probably at a relatively late stage in the deglaciation. The relics of this ice-stream track in the Tweed Basin are undoubtedly the best examples in the British Isles.

Striations

Where ice has moved over a horizontal or dipping exposed rock surface, the latter is often scarred and scraped by boulders within the moving ice. These parallel scratches clearly show the direction in which the ice has moved. Many fine examples are preserved along the East Lothian coast, often with overlying till, and these have been used effectively in determining precisely how the ice has flowed.

Glacial erratics

Erratics are boulders, often of large size, which have been carried by the ice and now rest on a floor formed by a different kind of rock. Their great value to glaciology is that, like the striations, they can show the directions of ice movement. Boulders of a distinct and recognisable type of rock often form trains leading outwards from a known point of origin. Tracking these boulders gives useful information about the direction in which the ice moved. Well-known examples include the southerly transport of the unmistakable riebeckite microgranite from Ailsa Craig in the Clyde estuary, but those in the Borders show that ice-flow was mainly eastwards.

Sea-level changes and isostasy

After the decay of the last (Devensian) ice sheets, two related events took place. The first was a global rise in sea-level as a direct result of the melting of huge volumes of ice. The second was what is known as isostatic rebound, in other words the overall rise of the land surface in response to the removal of the ice. The greater the original thickness of the ice and the greater the weight removed, the more extreme is the rebound. Isostatic uplift will raise the level of the land, while sea-level rise will drown it. Where the results of isostasy are greater than those related to rising sea-levels, then original beaches and wave-cut platforms become raised above the present sea-level, to be left stranded on land. This is very much the case in eastern Scotland and in the Borders, where the ice was thick. Where, however, there was little or no ice (paradoxically in Orkney and Shetland, which were further away from the centre of ice build-up) then some submergence of the land took place.

A very obvious example can be seen when travelling southwards along the Berwickshire coast. Some distance north of Berwick the railway line (and the A1 road) follows the top of a steep slope that was originally a sea-cliff. Below this is an area of almost flat, outwardly sloping, green farmland from which another cliff falls into the sea. This stretch of farmland was a former wave-cut platform, which lay below sea-level and was planed by wave action. Isostatic uplift has raised it up to 20 m above the present high-tide mark.

Raised beaches may be seen all around the East Lothian coast: at least five main ones have been identified, the earliest having formed between 16,000 and 14,000 years BP, the most recent about 9500 BP; they are higher inland where the ice was thicker.

The isostatic uplift of land results in down-cutting by rivers. Where there is an established floodplain, the river, rejuvenated by upward movement of the land, cuts down through its own deposits, and often into bedrock. The original floodplain is left above the new level, and as a result a pair of river terraces is formed with steep banks leading down to the new level. The valley of Kilbucho (Chapter 11), south of Biggar and between White Hill and Culter Fell, is actually a watershed from which one small river flows westwards and another to the east (Fig. 16.6). The eastward-flowing river, the Mitchelhill Burn, has carved three successive river terraces (Fig. 16.7) which must have been the result of pulsed uplift. The fertile soil of the original floodplain, now the uppermost terrace upon which the ruined church of St Bega stands, is surely one reason why this valley has been inhabited for at least two thousand years.

16.6 The valley of Kilbucho, looking westwards. Culter Fell to the left.

16.7 Three successive river terraces at Kilbucho, south of Biggar.

A closer look at the last 120,000 years

We have noted that the most recent glacial episode, knows as the Devensian, began about 120,000 years ago. In the early Devensian it seems that there were rapid climatic changes, although it was nothing like as cold as it became later. Stadial periods, i.e. those with growth of glaciers if there was enough precipitation, alternated with interstadial periods when the glaciers decayed. It was never warm enough, however, to allow a temperate interglacial. In the middle Devensian, large areas of Scotland may have been ice-free, as the bones of mammoth, reindeer, and woolly rhinoceros would suggest; these are some 30,000 years old.

But this was not to last long. Soon afterwards it became much colder. Ice sheets spread from the Highlands and the Southern Uplands, eventually merging with the European ice sheet centred on Scandinavia, and, further south, the ice from Ireland. There is some evidence that the Highland ice formed first and extended across the Midland Valley at about 26,000 years BP, before coalescing with the growing ice cap forming on the Southern Uplands. At its maximum extent the British Devensian ice sheet extended as far south as Wales and Yorkshire, but not as far as the ice sheet of the preceding (Anglian) glaciation which reached as far south as the Thames Valley. South of the ice sheet there was a cold wasteland with very little vegetation.

The intense cold of the later Devensian lasted for several thousand years. Yet by about 15,000 years BP, the ice slowly began to retreat round its margins and by about 14,600 years BP the climate was ameliorating rapidly enough to cause a fairly uninterrupted decay of the ice sheet. But then, from about 12,800 years BP, there came a further (though less severe) cooling event that lasted for about a thousand years.

16.8 Extent of the Loch Lomond readvance. Solid ice shown as diagonal stripes, glaciated high tops in black.

The ice came back, but only in the western Highlands. This was the time of the Loch Lomond Stadial, or Loch Lomond Re-advance as it is sometimes known. Figure 16.8 shows the maximum extent of the ice sheet, as indicated by its terminal moraines. Whereas this covered nothing like the extent of the great European ice sheets of 20,000 BP, the temperature has been estimated as some 10–12°C lower than it is at present. It is probable that the highest mountains in the Borders were glaciated during this time, but it was never an important area of spreading ice.

Around 11,500 BP came a dramatic period of global warming, the last ice melted and the Loch Lomond Stadial came to a rapid end. The climate flipped, and the temperature jumped some 10°C in a very short time (estimates range from twenty to seventy years). We can imagine the immediate effect of this fast sea-level rise, with inundation of low-lying coastal plains accompanied by immense winds. Thereafter the temperature rise was slower, ushering in the Flandrian Interglacial in which we now live.

The mild climate of the Flandrian interglacial reached an optimum around 5,000 BP, since when there has been a gradual global deterioration. There were evident global effects, on a larger scale; some 5000 years BP what is now the Sahara desert was a green wetland, and the same was probably true for Central Asia.

How can past climatic changes be tracked?

Cores which have been made through the Greenland ice cap have provided a detailed record of fluctuations in O^{16} and O^{18} ratios, which can be used to track temperature changes, as also can oxygen isotope ratios in marine foraminiferans. Similarly, various marine bivalve shells can be used, as those adapted to cold Arctic waters are quite distinct from those of more temperate seas, and the presence of such Arctic shells at particular horizons within a Quaternary marine sequence gives an unequivocal signal. All these different kinds of data tell a consistent story; a temperature rise at about 14,500 BP, followed by a slow, pulsed decline, and until about 11,500 BP, when there was a rapid jump, followed by a slower rise. Moreover, this consistency is complemented and refined by what at first sight seems to be an unlikely land-based source. Beetles!

The beetles living in Europe today are diverse and widespread. Most species are easily recognisable, and their remains preserve well in ancient soil and peat. Fossil beetles, belonging to the same species as those of today, can be easily identified, and the pupae in particular are of great value. Many living species of beetles have a restricted habitat and temperature preferences; the northerly cold-adapted assemblages are quite different from those of warmer climates. An alternation of warm- and cold-loving beetles through a peat sequence can tell, often with great precision, the relative temperatures at which particular horizons were formed. Recently the remains of chironomids (non-biting midges) have been used for tracking climate changes. This approach, pioneered during the 1960s by Russell Coope of Birmingham University, has been eminently successful, and research in this field continues at the present time at Edinburgh and elsewhere.

Pollen analysis

In Scotland, as elsewhere, changes in the vegetation since the last glaciers vanished can be readily assessed by pollen analysis. Pollen grains are normally easily identifiable: they do not readily decay, and they preserve a permanent record of what kind of vegetation existed at any particular time. In order to obtain useful data, a vertical core or section is taken through a peatbog, lake deposit, or soil. The pollen grains from each closely spaced level are extracted, identified and counted in ascending order. Whereas this is a time-consuming process, the end result will be a pollen spectrum, covering several thousand years, which shows not only the presence but the abundance and proportion of particular trees and other plants for each time period. Actual dates can be fixed to the pollen spectrum by radiocarbon dating.

We have seen that some 15,000 years ago there was a dramatic warming which led to average July temperatures, in this Late Glacial Interstadial, of some 15°, much the same as it is today. The initially open ground was rapidly colonised by tundra communities, with little alpine flowers such as *Dryas octopetala* and saxifrages. But these were soon replaced by grassland, heathers, and also willow (*Salix*) and juniper (*Juniperus*) scrubland; this kind of plant community is remarkably uniform throughout the British Isles. It seems unlikely that the scattered trees and shrubs formed a closed canopy. A change to grassland once more heralded the much colder conditions of the short-lived but intense Loch Lomond Stadial, with its attendant glaciers in the western Highlands of Scotland.

When the last ice retreated, the kind of vegetation that took over the scene was remarkably like that of the early part of the Late Glacial Interstadial, with juniper and willow and attendant vegetation. But these soon declined, very probably as a result of shading by large trees, which were to become very abundant, and formed true canopies. Of these, the first was the birch (*Betula*), at about 10,800 to 10,600 BP, and it may have migrated from unglaciated refuges in the west. Hazel (*Corylus*) appeared a few hundred years later. Scots Pine (*Pinus sylvestris*) established itself in western parts of Scotland, but in the Borders does not seem to have been able to colonise the ground successfully. It may be that the soil was too waterlogged, and thereafter pine seems not to have been able to compete, in this area, with broadleaved trees. There followed oak (*Quercus*), and elm (*Ulmus*) at around 8,000 BP or earlier in some places, especially at lower altitudes. Alder (*Alnus*) became locally abundant, evidently being able to colonise small specific sites before expanding. Lime (*Tilia*), which prefers warmer conditions, became fairly common in England around 6000 BP, but only just came into the Borders, in the Cheviot region. The hills in the Borders by this stage were covered with forest, and it is highly probable that even the highest summits were forested.

The major changes that took place from about 6000 BP onwards can be directly attributed to human impact. In particular, elm began to decline after about 5800 BP, which at least to some extent can probably be attributed to forest clearance. The lowlands became agriculturally productive, and by the early Bronze Age, around 4500 BP, farmers began to move into more upland regions. The destruction of woodland accelerated, and by the time of the Roman occupation there may actually have been less woodland than there is now. Thereafter, felling, burning and the spread of agriculture in mediaeval times led to the almost total destruction of the original forest.

16.9 Glaciated Ordovician hills, near Coulter, south of Biggar.

Although the ancient woodland has gone, we now have the possibility of seeing what it was like. In 1999 the Borders Forest Trust bought 600 hectares of land on the hillside at Carrifran, in the ice-carved valley east of Moffat, and planted a fine woodland of the kind of trees that used to be there; birch, hazel, alder and blackthorn. The Carrifran Wildwood, as it is known, is a Scottish treasure, not only for the present time, but for generations to come.

Finally, consider the striking photograph (Fig. 16.9) of Ordovician rocks of the Northern Belt, glaciated some thousands of years ago, their structures picked out by the sunlight of late October. The ancient ocean, in which they formed, has long gone. But the timeless beauty and fascination of these hills remains.

Select bibliography

Armstrong. H. and Brasier, M. (2005) *Microfossils* (2nd edition). Blackwell Science.

Benton, M.J. (2003) *Vertebrate Palaeontology* (3rd edition). Blackwell Publishing.

Cameron, I.B., and Stephenson. D. (eds) (1985) *The Midland Valley of Scotland:British Regional Geology* (3rd edition). British Geological Survey.

Clarkson, E.N.K. (1998) *Invertebrate Palaeontology and Evolution* (4th edition). Blackwell Science.

Clarkson, E. and Upton, B. (2006) *Edinburgh Rock: the Geology of Lothian.* Dunedin Academic Press.

Grieg, D.C. (1971) *The South of Scotland: British Regional Geology* (3rd edition). Institute of Geological Science.

McAdam, A.D. and Clarkson, E.N.K. (1986) *Lothian Geology: an excursion guide.* Edinburgh Geological Society and Scottish Academic Press.

McAdam, A.D., Clarkson, E.N.K. and Stone, P. (1996) *Scottish Borders Geology; an excursion guide.* Edinburgh Geological Society and Scottish Academic Press.

Repcheck, J. (2003) *The man who found time: James Hutton and the discovery of the Earth's antiquity.* Pocket Books.

Sissons, J.B. (1976) *The Geomorphology of the British Isles: Scotland.* Methuen.

Trewin, N.H. (2002) *The Geology of Scotland* (4th edition). Geological Society of London Publishing House.

Trewin, N.H. (2008) *Fossils alive!* Dunedin Academic Press.

Upton, B.G.J. (2004) *Volcanoes and the making of Scotland.* Dunedin Academic Press.

Index